Reviews and critical articles covering the entire field of
normal anatomy (cytology, histology, cyto- and histochemistry, electron microscopy, macroscopy, experimental
morphology and embryology and comparative anatomy)
are published in Advances in Anatomy, Embryology and
Cell Biology. Papers dealing with anthropology and clinical
morphology that aim to encourage co-operation between
anatomy and related disciplines will also be accepted.
Papers are normally commissioned. Original papers and
communications may be submitted and will be considered
for publication provided they meet the requirements of a
review article and thus fit into the scope of "Advances".
English language is preferred, but in exceptional cases
French or German papers will be accepted.
It is a fundamental condition that submitted manuscripts
have not been and will not simultaneously be submitted or
published elsewhere. With the acceptance of a manuscript
for publication, the publisher acquires full and exclusive
copyright for all languages and countries.
Twenty-five copies of each paper are supplied free of charge.

Manuscripts should be addressed to

Prof. Dr. A. **BRODAL,** Universitetet i Oslo, Anatomisk Institutt,
Karl Johans Gate 47 (Domus Media), Oslo 1/Norway

Prof. W. **HILD,** Department of Anatomy, Medical Branch, The University
of Texas, Galveston, Texas 77550/USA

Prof. Dr. J. van **LIMBORGH,** Universiteit van Amsterdam, Anatomisch-
Embryologisch Laboratorium, Mauritskade 61, Amsterdam-O/Holland

Prof. Dr. R. **ORTMANN,** Anatomisches Institut der Universität, Lindenburg,
D-5000 Köln-Lindenthal

Prof. Dr. T. H. **SCHIEBLER,** Anatomisches Institut der Universität,
Koellikerstraße 6, D-8700 Würzburg

Prof. Dr. G. **TÖNDURY,** Direktion der Anatomie, Gloriastraße 19,
CH-8006 Zürich/Schweiz

Prof. Dr. E. **WOLFF,** Collège de France, Laboratoire d'Embryologie
Expérimentale, 11, Place Marcelin Berthelot, F-75005 Paris/France

Advances in Anatomy
Embryology and Cell Biology

Vol. 61

Editors
A. Brodal, Oslo   W. Hild, Galveston
J. van Limborgh, Amsterdam
R. Ortmann, Köln   T.H. Schiebler, Würzburg
G. Töndury, Zürich   E. Wolff, Paris

Hubert Korr

Proliferation
of Different Cell Types
in the Brain

With 21 Figures

Springer-Verlag
Berlin Heidelberg New York 1980

Dr. rer. nat. Dr. med. habil. Hubert Korr
Institut für Medizinische Strahlenkunde der Universität
Versbacher Straße 5
D-8700 Würzburg

ISBN-13:978-3-540-09899-7   e-ISBN-13: 978-3-642-67577-5
DOI: 10.1007/978-3-642-67577-5

Library of Congress Cataloging in Publication Data. Korr, H. 1941- Profi-
leration of different cell types in the brain. (Advances in anatomy,
embryology, and cell biology; v. 61) Bibliography: p. Includes index.
1. Neurons. 2. Cell proliferation. 3. Brain-Cytology. I. Title. II. Series.
[DNLM: 1. Brain Cytology. 2. Cell division. W1 AD433K v. 61 WL300
K845p] QL801.E67 vol. 61 [QL931] 574.4s [599.01'88] 80-11403

Composition: SatzStudio Pfeifer, Germering

Contents

Acknowledgments

I would like to thank Professor A.M. Kellerer, Professor W. Maurer, and, particularly, Professor Brigitte Maurer-Schultze for their advice and criticism during the preparation of the manuscript. The careful translation of T.C. Telger is gratefully acknowledged.

1 Introduction

Studies on cell kinetics in untreated animals have for the most part been done on organs in which many proliferating cells can be found. In general the proliferating cells have been identified either in histologic sections as mitoses or by autoradiography as labeled interphase cells following the injection of a labeled precursor of DNA, such as ^{3}H- or ^{14}C-thymidine (TdR).

A great many proliferating cells can be observed in the rat and mouse brain during the embryonic period and for a short time after birth, and many studies on cell kinetics have been performed for this phase of life. By contrast, very few proliferating cells are found in the brain of *adult* rodents (except for the subependymal layer, see below). As a result, only isolated studies have been done on cell kinetics during this period. Although there is an increase in proliferating cells in adult animals which had been pretreated (e g. , by wounding, X-irradiation, viral infection, withdrawal of water), this proliferation too has not been investigated in detail.

A number of studies have been done since 1959 on the proliferation of *cells in the sub-ependymal layer* of the lateral ventricles of the forebrain. This cell type is well suited for such investigations because mitoses can be found there even in animals which are quite old.

Since the studies of Leblond and co-workers (Walker and Leblond 1958; Messier et al. 1958), it has been known that some of the *glial cells* in the brain of adult mice can be labeled by administering labeled TdR. However, it was unclear whether this labeling after injection of ^{3}H-TdR was really an expression of DNA synthesis within a cell cycle with subsequent mitosis. One reason was that only a small number of labeled glial cells were found in short term experiments (e. g., only about 2 of 1000 glial cells in the forebrain of the untreated adult mouse). Furthermore, hardly any mitoses could be detected. The absence of mitotic figures led many to speculate well into the 1960s on an amitotic mechanism for glial cell division (Hansson and Sourander 1964; Hugosson et al. 1968). However, Fleischhauer (1967) and Cavanagh and Lewis (1969) showed that the absence of mitoses in the CNS was due to a fixation artifact: Experiments with perfusion and immersion fixation clearly showed that the detection of mitoses in the CNS depends upon rapid penetration of the fixation fluid. When perfusion fixation was done, glial cell mitoses could be regularly observed.

Based on the number of labeled glial cells found in brain sections and the observation of few labeled mitoses, the first quantitative discussions on glial cell proliferation were published in 1961. Smart and Leblond (1961), for example, estimated that about 150 000 to 240 000 new glial cells are formed per day in the mouse brain. Parallel DNA determinations in the mouse brain showed no increase of DNA with age. These findings led the authors to conclude that an equilibrium must exist between the production of new glial cells and cell loss. Dalton et al. (1968) reached a similar conclusion: From their measurements of the percentages of labeled glial cells in the mouse forebrain as a function of age, they calculated a cell production rate of 0.59% to 0.24% per day. These values were then compared with published data on the increase in brain weight with age. From this comparison, it was found that more glial cells were produced than could be accounted for by the increase in brain weight. Taken as a whole, these initial estimates were sufficient to indicate that the glial cell population was not

1

static, as had been previously assumed (Leblond and Walker 1956), but probably underwent a gradual renewal.

Experimentally determined cell cycle parameters such as the length of the S phase (T_S) or the cell cycle time (T_C) were not available for the glial cells until 1973. Rough estimates such as the cell production rate mentioned above were based on unproved assumptions on the length of the S phase. Noetzel (1962) calculated a cell cycle time of about 50 days for the glial cells in the brain of adult mice based on his measurements of the labeling index (0.65%) and an assumed S phase length of 7 h. It is assumed in such a calculation that all glial cells take part in proliferation (growth fraction = 1), an assumption which is entirely hypothetical.

Before 1973, nothing was known about the proliferation of *other cell types* in the rodent brain (endothelial cells, microglia and pericytes, cells of the ependyma and choroid plexus, the meningeal cells or the neurons). Only a few reports had been published on the labeling of certain cell types after injection of labeled TdR in adult animals. However, these isolated findings were contradictory in many cases and offered an uncertain basis for judging which of the aforementioned cell types in the brain of untreated young and adult rodents could actually be labeled with TdR.

With the fragmentary nature of knowledge in this area, it appeared necessary to undertake more detailed investigations on the proliferation of different cell types in the rat and mouse CNS. Our own investigations, in progress since 1971, were concerned initially with the proliferation of neuroglia and endothelial cells in the forebrain of adult mice. The cell cycle and mode of proliferation of these cells were analyzed by means of double-labeling experiments with ^{3}H- and ^{14}C-TdR and two-emulsion-layer autoradiography (Korr 1973, 1974, 1978a; Korr et al. 1973, 1975, to be published a, d). Later these investigations were extended to include glial and endothelial cells in four regions of the forebrain of 14-day-old rats, i.e., animals which had not yet reached adulthood (Korr 1978a; Korr et al., to be published b, c). With the aid of a specially developed method (combination of metallic impregnation and autoradiography: Korr 1978b), it could be demonstrated that only astrocytes and oligodendrocytes but not microglial cells proliferate in the brain of the untreated adult mouse. This made it possible for the first time to measure the length of the cell cycle phases for astrocytes and oligodendrocytes. A surprisingly short cell cycle time of about 20 h was found for both types of glial cells in both the adult mouse and the young rat. A length of about 10 h was measured for the DNA synthesis phase. The length of the phases (G_2 + M) was estimated at 4-5 h, and the duration of the G_1 phase at 5-6 h. Moreover, a correlation could be established between glial cell proliferation and the occurence of pyknotic glial cells: Glial cell proliferation in the untreated animal is accompanied by a constant cell loss, this loss being greater in adult animals than in young. Furthermore, it could be shown by various autoradiographic studies that a constant exchange takes place between the pool of proliferating glial cells and the pool of nonproliferating glia.

Further studies dealt with the period during which microglia, pericytes, ependymal cells, and epithelial cells of the choroid plexus proliferate (Korr 1978 a). It was found that these cell types could be labeled not only prenatally, but also postnatally in the first weeks of life. Proliferating ependymal cells could be observed sporadically in adult animals as well.

Based on our results in young as well as in old mice and rats, data already published by other authors, especially studies on embryonic brains, could be viewed in a new context. It became possible to discern the basic features of the proliferation of different cell types (Korr 1978a, cf. also Schultze-Maurer 1978). For example, it was found that cell cycle parameters previously known for different cell types in the rodent brain largely coincide and remain more or less constant from the end of the fetal period throughout postnatal life. The proliferation with cell loss observed for neuroglial cells, subependymal cells, and endothelial cells in the brain of young and adult rodents appears also to be present prenatally in the neuroepithelial cells, i. e., the precursor cells of the neuroglia, neurons, and ependyma. Evidence was also found to indicate that the exchange processes between proliferating and nonproliferating cells found in young and adult animals, particularly in the neuroglia, also predominate prenatally in the neuroepithelial cells.

It is the object of the present work to answer the following questions on the proliferation of different cell types in the brain of untreated rats and mice during the course of pre- and postnatal ontogeny, based on the current state of knowledge:

1. In what period of ontogeny do the individual cell types or their precursors proliferate?

2. Which are the cell cycle parameters of these cell types?

3. What can be concluded about the mode of proliferation of the individual cell types?

As these questions are answered, the special autoradiographic methods used to resolve them will also be discussed.

2 Thymidine Labeling as an Indication of Cell Proliferation

Before discussing experiments with labeled TdR on different cell types in the CNS, we should first address the question of whether labeling after administration of labeled TdR is really an expression of cell proliferation.

Ever since the studies of Reichard and Estborn (1951) with ^{15}N-labeled TdR, it has been confirmed by many authors that TdR is a specific precursor of DNA (e.g., Feinendegen 1967, Schultze 1969). While it is true that Bryant (1966) and Lima-de-Faria (1965) showed that tritium can also be found in protein fractions after injection of ^{3}H-TdR, the degree of protein labeling is too slight to preclude the use of TdR as a specific DNA precursor (cf. also Goldspink and Goldberg 1973).

In the mammalian cell, the DNA is localized primarily in the cell nucleus, but can also be found in the mitochondria of the cytoplasm (DNA content of the mitochondria is ca. 0.15% of the DNA content of the cell nucleus [Nass 1969]). According to studies by various authors (survey: e.g., Schultze 1969), this mitochondrial DNA is also capable of synthesizing new DNA. However, a comparison of DNA synthesis in the nucleus and the mitochondria shows that the labeling of mitochondrial DNA is so weak that it does not affect the question of whether a cell is labeled or not.

In their experiments with ^{32}P on root meristems of Vicia faba, Howard and Pelc (1949, 1950, 1951) showed that DNA is synthesized only by cells which are preparing for division. Moreover, these authors postulated even then that DNA synthesis is restricted to a specific time period within the interphase. This hypothesis has subsequently been confirmed for numerous somatic cells (survey: e.g., Schultze 1969).

It has been suggested in most previous works that the labeling of brain cells in young or adult mammals after injection of labeled TdR is the result of a DNA synthesis phase within the cell cycle. A simple way to confirm this experimentally is to measure the percentage of labeled mitoses as a function of time following a single injection of labeled TdR (PLM method, cf. Quastler and Sherman 1959). In this method the passage of the originally labeled S phase cells through the next and subsequent mitoses is recorded, thereby establishing an identity between the labeled cells and proliferating cells. Such investigations have previously been done in the adult mammalian brain only for cells of the subependymal layer (Lewis 1968a; Gracheva 1969; Lewis and Lai 1974; Lewis et al. 1977b). Since mitoses in the adult animal brain are very rarely observed outside the subependymal layer, Korr et al. (1973) took a different approach to demonstrate experimentally the connection between the uptake of labeled TdR and cell proliferation. In contrast to the technique mentioned above, i. e., using the PLM method to show the existence of the S phase and measure its length, Korr et al. (1973) did studies on labeled interphase cells. Double-labeling experiments with ^{3}H- and ^{14}C-TdR showed a length of about 10 h for the DNA synthesis phase of glial and endothelial cells in the adult mouse brain. This period is similar to the S phase length in the cell cycle of many other somatic cells in various animals (cf. Maurer and Schultze 1968). The further observation that only unlabeled glial and endothelial cell mitoses were observed initially after injection of ^{3}H-TdR, but labeled mitoses appeared after a period of about 3 h, then made it likely that a specific, circumscribed DNA synthesis phase actually exists within the cell cycle for glial and endothelial cells.

Now the presence of labeled mitotic figures, especially prophases, does not necessarily prove that cell division has actually occurred, and thus that a normal cell cycle exists. It is known, for example, that mitosis in hepatocytes can reverse itself as long as chromatid separation is not complete (cf. Pera 1970; Schultze et al. 1973; Gerhard et al. 1973). This results in a polyploid cell. However, polyploidy in the brain has been observed only in a few neuron groups (survey: Lapham et al. 1971; Jacobson 1978), although these findings are not undisputed (cf. Mann and Yates 1973a,b).

At present it is generally assumed that DNA, as the carrier of genetic information, is metabolically stable. This implies that a cell labeled with labeled TdR will retain its label as long as the content of labeled DNA does not fall below detectable limits as the cell divisions proceed. This conclusion has been confirmed experimentally for cells of the CNS by Mares et al. (1974): The authors were able to show that Purkinje cells in the mouse cerebellum labeled prenatally with ^{3}H-TdR retained a constant ^{3}H-activity over a period of 90 days after birth. This study also demonstrated that the conclusions of Haas et al. (1970), who had assumed an instability of neuronal DNA based on autoradiographic findings in various neurons of the neocortex, were based on a misinterpretation of experimental data.

The assumption of DNA stability is not uncontested, however. Pelc, in particular, postulated the existence of a "metabolic DNA" in addition to stable DNA (survey: e.g., Pelc 1970, 1972; cf. also Roels 1966, Schultze 1969; Adrian 1971). If true, this would mean that labeling with ^{3}H- or ^{14}C-TdR could no longer be considered evidence of cell proliferation in every case.

Pelc cites various examples in his works and concludes the existence of a metabolic DNA based on a combination of various experimental values. In the case of the crypt epithelial cells of the mouse jejunum, however, it was shown experimentally by Maurer et al. (1972) and Burholt et al. (1973) that the conclusions of Pelc were based on false premises. Moreover, the aforementioned investigations of Mares et al. (1974) on Purkinje cells from the mouse cerebellum produced no evidence to indicate the existence of a metabolic DNA.

Despite the convincing arguments against the existence of a metabolic DNA, the DNA molecule can no longer be considered absolutely stable. It has been known since about 1964, owing to investigations by the Howard-Flanders and Setlow groups, that the cell replaces damaged DNA segments (survey on DNA repair: e.g., Howard-Flanders 1968; Strauss 1968, 1974; Painter 1970; Hanawalt 1972; Cerutti 1974). The incorporation of labeled TdR outside the S phase ("unscheduled DNA synthesis") has been demonstrated, particularly in cell cultures, following irradiation, treatment with certain drugs, and even "spontaneously" (see survey works on DNA repair listed above). Such DNA repair processes are also known for cell types of the CNS: Wheeler and Lett (1972), after exposing the cerebellum of young dogs to X-rays, found that strand breaks produced in the DNA of neurons were repaired in vivo. In untreated specimens, on the other hand, the incorporation of labeled TdR outside the S phase has been observed only in Hela cells in vitro (Djordjevic et al. 1969; Evans et al. 1970). The resulting grain counts per nucleus were very low. It can be concluded, then, that this phenomenon − if it exists at all in the CNS cell types in vivo − will cause no confusion with DNA synthesis during the S phase.

Investigations by Adelstein et al. (1964) in various rodent species should also be mentioned. These authors showed that cells which were unlabeled after administration of labeled TdR were not necessarily outside the S phase. In various North American squirrel species, for example, proliferating cells could not be labeled with ^{3}H-TdR. Biochemical studies made it clear that certain (unspecified) defects on a cellular level interfered with the uptake of exogenously administered TdR.

3 Period of Proliferation of Different Cell Types in the Rat and Mouse Brain

The present chapter deals with the periods during pre- and postnatal ontogeny in which the individual cell types or their direct precursors are capable of proliferation. The degree of proliferation, i.e., the question of how many cells of a particular type are proliferating, will not be discussed here.

3.1 Methods

In *postnatal* life the different cell types can be readily differentiated histologically, and it is not difficult to determine the end of the proliferation phase of a particular cell type. Labeled ^{3}H- or ^{14}C- TdR is injected, and the animal is killed a short time (usually 1 h) later (short-term experiments). It can then be determined whether cells of a particular type were proliferating at the time of TdR injection. However, if the object is to learn whether a cell type develops from precursor cells, the experimental animal must be allowed to survive for some time (e.g., several weeks) after the TdR injection.

In *embryonic* brains (especially during the early embryonic period), it is often impossible to identifiy individual cell types histologically, even with the aid of the electron microscope. As a result, the proliferation of a particular cell type cannot be investigated in this period by short-term experiments with ^{3}H- or ^{14}C-TdR. It can be

5

studied experimentally only by labeling the fetal brain cells by maternal injection of ^{3}H-TdR at a specified point during embryonic development (e.g., at embryonic day 11, abbreviated E 11). The offspring of the ^{3}H-TdR-injected mothers are then killed postnatally at a time (e.g., at postnatal day 25, or P 25) when all cells in the brain can be clearly identified histologically. The detection of labeled cells in such an animal signifies first that these cells or their precursors were proliferating at the time of TdR injection, i.e., were in the S phase of a cell cycle. Persistence of the labeling, moreover, indicates that these cells underwent only a few divisions after prenatal ^{3}H-TdR injection. Cells which are heavily labeled must have stopped proliferating very soon after labeling, i.e., after about one or two mitoses; many cell divisions would have so diluted the label that it could no longer have been detected. The end of the proliferation phase of a particular cell type is deduced from the absence or presence of heavily labeled cells as a function of the time of the prenatal injection: The end of proliferation corresponds approximately to the time at which prenatal ^{3}H-TdR injection produces for the last time labeled cells of the type investigated in the young animal. Of course, this method of prenatal labeling and postnatal examination cannot be used to ascertain the time at which the cells start to proliferate, since the label is too highly diluted up to the end of proliferation (cf. Schultze et al. 1974a).

3.2 Period of Proliferation

3.2.1 Neurons

All types of neurons, as well as the neuroglial cells (astrocytes and oligodendrocytes), the cells of the subependymal layer, the ependymal cells, and the epithelial cells of the choroid plexus arise from a common precursor cell type, the neuroepithelial cells. Other synonyms for these cells are germinal cells (His 1889), ependymal cells (Schaper 1897a,b), primitive ependymal cells (Sidman et al. 1959), matrix cells (Fujita 1963), and ventricular cells (Boulder Committee 1970). This common precursor cell cannot yet be differentiated morphologically with regard to its later development (cf. Meller et al. 1966). Those neuroepithelial cells which later differentiate into neurons are called neuroblasts after their last mitotic division. The neuroblast then matures into a neuron.

Start of proliferation. According to a concept of Fujita (1965, 1966), the formation of the neural plate [at about E 7 in the mouse (Snell and Stevens 1966) and about E 9 in the rat (Hagemann and Schmidt 1960)] is followed first by "stage I": the production of neuroepithelial cells which are functionally homogenous. Following this stage (at an unspecified time) is "stage II": the production of the future neuroblasts. According to this theory, then, the precursor cells of the neuroblasts proliferate early in the embryonic period. For the reasons mentioned above, the time at which this proliferation occurs cannot be measured experimentally.

Nevertheless, Schultze et al. (1974 a,b) were able to determine indirectly the time at which neuroepithelial cells which later differentiate into neurons of specific nuclei in the rat brain (e.g., the neurons of the n. ruber) start to proliferate. In these investigations the rats received ^{3}H-TdR at various times during embryonic development, The brains of the young animals were studied postnatally (at P 25) by autoradiography. The autoradiographs were evaluated quantitatively: The percentage of labeled neurons was measured in specified regions of the brain (e.g., the n. ruber) as a function of the time of ^{3}H-TdR injection, and the number of silver grains per cell nucleus were counted. The key insight was that the percentage of labeled neurons found at P 25 corresponds to the labeling index of the corresponding neuroepithelial cells at the time of prenatal ^{3}H-TdR injection, which otherwise cannot be measured experimentally. Thus, neurons labeling which could be measured postnatally provided some information on the prenatal proliferation of the precursor cells.

Knowing the labeling index and the length of the S phase (taken from data in the literature), Schultze et al. (1974 a,b) were further able to calculate the cycle time of those neuroepithelial cells from which a particular type of nerve cell (i.e., the neurons of a brain nucleus) later arises by differentiation. By about day 16 of fetal life, a more or less consistent cycle time of about 10 h was found for the investigated neurons of the supraoptic nucleus, n. ruber and n. cochlearis ventralis, in the piriform cortex, for the hippocampal pyramidal cells, for the cerebellar Purkinje cells, and for the neurons in neocortex layers II and III. Accordingly, the precursors of these neuron types proliferate initially as a homogenous cell population.

The end of proliferation occurred at various times for the neuron types indicated: It ranged from embryonic day 13 (n. ruber) to about fetal day 21 (hippocampal pyramidal cells). From the knowledge of the end of proliferation, the cell cycle time, and from the histometrically determined number of neurons in a particular brain region, the number of mitotic divisions could be estimated and thus what period of time was necessary to produce the corresponding number of cells measured. It was thus possible to determine the time at which the precursor cells had started to proliferate. Schultze et al. (1974a,b) found a uniform onset of neuroepithelial cell proliferation for four different types of nerve cell (i.e., for those investigated nuclei for which the number of neurons was known): about the 9th day of embryonic development. This coincides with the formation of the neural plate in the rat embryo.

End of proliferation. For the precursor cells of most neurons, the proliferation phase ends prenatally. This point in time, which is also called the "date of birth," "cell birthday," or "time of origin" of a neuron group, has been determined autoradiographically for many neuron groups (survey: e.g., Altman 1969a; Angevine 1970; Sidman 1970; Langman et al. 1971; Jacobson 1978 (including a complete bibliography up to about 1975); see also Brückner et al. 1976; Rickmann et al. 1977; Lawson et al. 1977; McAllister and Das 1977; Shimada et al. 1977; Sturrock 1977, 1978b; Altman and Bayer 1977, 1978a,b,c; Schlessinger et al. 1978; Anderson 1978; Raedler and Raedler 1978; Nornes and Carry 1978; Bayer 1979; Sims and Vaughn 1979). However, some precursor cells of neurons, especially cerebellar neurons, continue to proliferate postnatally during about the first 3 weeks of life (survey: e.g., Altman 1970).

It would appear that new neurons are no longer formed in the brain of *adult* rodents, according to the results of Messier and Leblond (1960). Schultze and Oehlert (1960), Smart and Leblond (1961), Adrian and Walker (1962), Walker (1963), Noetzel and Rox (1964), Dill and Walker (1966), Kraus-Ruppert et al. (1973), Mares et al. (1974, 1975), and Korr et al. (1973, 1975). These authors state explicitly that they found no labeled neurons after the injection of ^{3}H-TdR, either in short-term experiments or in long-term experiments up to 30 days after injection. Manuelidis and Manuelidis (1971) also reported finding no labeled nerve cells in their in vitro experiments.

7

Contrary findings were reported by Messier et al. (1958), Altman (1962a,b, 1963), Pelc (1972), Reinis (1972) and recently by Kaplan and Hinds (1977). Altman described labeled neurons 1 day to 2 months after ^{3}H-TdR injection following the production of brain lesions (Altman 1962a,b); but even in untreated adult rats he observed labeled neurons 2 weeks after ^{3}H-TdR injection (Altman 1963). However, the author himself stressed the difficulties of identifying with certainty a labeled cell as a neuron. On the one hand, this is true for the small neurons in the cerebellar granular layer as well as in the hippocampal dentate gyrus. On the other hand, a labeled glial cell which is in close proximity to the neuron ("perineuronal satellite glia") can create the impression, depending on the geometry of the section, that the neuron itself is labeled. Even after considering these difficulties, Altman (1963) still spoke of labeled neurons. It is noteworthy, however, that in many examples of "labeled neurons" the cell nucleus is not labeled homogenously (as is customarily the case in many other cell types), but shows silver grains only over a more or less circular segment of the nucleus. In his paper of 1963, Altman also reported neurons having two circular labeled segments. This is apparently the case where labeled pairs of glial cells were in a satellite configuration − an observation also made repeatedly in our studies on adult mice.

The figures of Messier et al. (1958) and Reinis (1972) showing "labeled neurons" are not convincing owing to the poor quality of the autoradiographs. In the case of Reinis (1972), moreover, the labeling appears to stem from glial cells. In the figure published by Pelc (1972), it is doubtful whether the labeled cell is a neuron at all.

Altman (1967, 1969a) presumed that even in the adult animal, small neurons with short axons ("microneurons": McLardy 1963: Altman and Das 1965) are formed into the olfactory bulb and hippocampus from cells migrated from the subependymal layer. In the case of the olfactory bulb, this assumption was substantiated by studies in young rats injected with ^{3}H-TdR at age 13 days (Altman 1966) or 30 days (Altman 1969b). A few labeled granular cells were observed in the hippocampal dentate gyrus even when 180- or 240-day-old rats had received ^{3}H-TdR and were killed 2 weeks after injection (Altman and Das 1965).

As mentioned above, it must be borne in mind in these studies that microneurons bear a close resemblance to glial cells (astrocytes) in the paraffin section.

Kaplan and Hinds (1977) reported on labeled neurons (granular cells) in the hippocampal dentate gyrus and in the olfactory bulb 30 days after injection of ^{3}H-TdR in 90-day-old rats. These authors identified cells found in light microscopic autoradiographs with the aid of the electron microscope after the 1-μm thick autoradiograph section had been reembedded and areas with labeled cells had been processed into ultrathin sections. Despite this first indication of possible neurogenesis in the adult animal, however, further evidence is necessary before the classic concept that no new nerve cells are formed in the adult mammalian brain can be abandoned. For example, it must be confirmed experimentally that the labeling in this case is actually attributable to DNA synthesis of the precursor cells within a cell cycle, rather than to polyploidization of local neurons (see Sect. 2). It would also be important to demonstrate that the number of labeled neurons increase with time after injection of ^{3}H-TdR.

3.2.2 Neuroglia (Astrocytes, Oligodendrocytes)

Start of proliferation. In his concept of matrix cells (see Sect. 3.2.1), Fujita (1965, 1966) assumed that in prenatal life the completion of the production of neuroepithelial cells which later become neurons (stage II) is followed by a "stage III" in which only those neuroepithelial cells proliferate which finally differentiate into neuroglial cells or into other cell species which also stem from the common precursor cell type. The start of stage III is believed to vary from one brain region to the next.

If this hypothesis is correct, the precursor cells of the neuroglia [usually called "glioblasts" (see Jacobson 1978), but sometimes called "spongioblasts" (His 1887)], probably could not start to proliferate very soon after formation of the neural plate during the early embryonic period.

Although, as stated earlier, it is not possible to determine experimentally the time of the start of proliferation (see Sect. 3.1), it is possible to obtain at least a rough estimate of it from the results of autoradiographic studies in which ^{3}H-TdR is injected prenatally and the animals are killed postnatally, i.e., it is possible to determine the time *before* which proliferation must start in the various regions of the brain. Hicks and D'Amato (1968) reported on heavily labeled glial cells in certain regions of the rat isocortex when the ^{3}H-TdR had been injected at gestational day 21 (E 21). Berry and Rogers (1966) found labeled glial cells in the cortex of young rats after injection at E 18. Similarly Schultze et al. (1974a) observed labeled glial cells in the corpus callosum of young rats if ^{3}H-TdR was injected at E 18/19. According to studies by Biesold et al. (1976), labeled glial cells are found in the dorsal nucleus of the lateral geniculate body of the rat following injection at E 14. Hinds (1968) reported finding labeled glial cells in the mouse olfactory bulb when ^{3}H-TdR was injected as early as E 12. Labeled glial cells were also found by Korr (1978a) in the n. ruber of the rat when ^{3}H-TdR was administered at E 11/12. Finally, Sturrock (1978b) found the first labeled glial cells in the mouse indusium griseum when the label was injected at E 11.

Various studies have been done to determine the type of glial cells which are labeled postnatally following a prenatal injection of ^{3}H-TdR. Berry and Rogers (1966) reported labeled astrocytes and oligodendrocytes as well as microglial cells in the rat cortex (^{3}H-TdR injected at E 18). They distinguished between the different cell types solely on the basis of the morphology of the cell nucleus in the paraffin section – a method which is not reliable (cf. Mori and Leblond 1969a,b, 1970; Ling et al. 1973). According to Skoff et al. (1976b), labeled astrocytes were found in electron microscopic autoradiographic studies of the rat optic nerve when ^{3}H-TdR was injected at fetal day 15 or 16. In this case labeled oligodendrocytes could be observed only when ^{3}H-TdR was injected postnatally.

Sturrock (1974c) used electron microscopy (without autoradiography) to investigate the time of origin of various glial cell types within the anterior commissure of the prenatal mouse brain. He observed glioblasts, which in his opinion would later develop into either oligodendrocytes or astrocytes, in the interval from E 16 to P 14. Oligodendroblasts and astroblasts were found in this commissure from E 17 to about P 32. These observations suggest that the direct precursor cells of oligodendrocytes and astrocytes may have started to proliferate before E 16 or 17.

Rickmann and Wolff (1976a,b) injected rats with ^{3}H-TdR at embryonic day 13 and then used the electron microscope 4 days later to examine the heavily labeled cells found previously in the autoradiographs by light microscopy. These heavily labeled cells (i.e., cells which had stopped proliferating shortly after labeling) could be identified as astroblasts. It was also possible to trace these cells back morphologically to early E 13. At this time they were discernible in the marginal area of the undifferentiated ventricular zone.

These latter results, as well as those of Sturrock (1978b) for the indusium griseum and our results for the glial cells of the n. ruber (Korr 1978a), appear to contradict Fujita's concept of the matrix cells (see Fujita 1965, 1966). In these three cases, cells which were clearly neuroglial precursors were found in circumscribed areas at a time when the proliferation of prospective neuroblasts was predominant. Thus, Fujita's "stage III" extends far into stage II and perhaps cannot be separated from it at all.

Summarizing the results presented on the start of proliferation of neuroglial cells, it can be stated that the precursor cells of the astrocytes clearly start to proliferate before birth and in some regions perhaps even during the early embryonic period before E 12. The oligodendrocytes also start to proliferate prenatally in some regions of the brain. It is unclear, however, whether the precursors of these cells tend to proliferate as early prenatally as the precursors·of the astrocytes. In any event, the majority of both the astrocyte and oligodendrocyte precursors probably do not start to proliferate until shortly after birth (cf. Sturrock 1974c, 1978b; Skoff et al. 1976a,b; Das 1977; Mares and Brückner 1978).

End of proliferation. As many studies with ^{3}H-TdR have shown, neuroglial cells can be labeled even in the brain of adult rats and mice which have not been pretreated in any way. However, the number of labeled glial cells in the adult animal brain is very small (labeling index of about 0.2% in the forebrain of adult mice; cf. Korr et al. 1975) and decreases with age (Dalton et al. 1968). Still, a few labeled oligodendrocytes and astrocytes have been found even in the brain of 18-month-old NMRI mice, and thus in senile animals (Korr, to be published; cells labeled after injection of ^{14}C-TdR were identified by a combined technique of metallic impregnation and autoradiography; Korr 1978b). This means that, generally speaking, astrocytes and oligodendrocytes are capable of proliferation throughout the lifetime of the animals.

3.2.3 Cells of the Subependymal Layer

The subependymal layer can be identified in the lateral ventricles of the brain toward the end of the embryonic period as a collection of undifferentiated, mitotically active cells between the ventricular layer and intermediate zone. In the mouse, this layer is discernible after about E 14 (Langman and Welch 1967). The subependymal layer continues to exist after birth and can be observed even in old age, especially in rodents (survey: Fleischhauer 1972; Jacobson 1978).

The proliferation phase thus begins prenatally with the appearance of the subependymal layer and extends throughout life. Labeled cells could be found in this layer even in short-term experiments in senile mice (Korr, to be published).

3.2.4 Ependymal Cells

Start of proliferation. The following results were obtained by the aforementioned method of prenatal ^{3}H-TdR injection and postnatal examination: Berry and Rogers (1966) found labeled ependymal cells in the brain of 30-day-old rats (probably in the lateral verntricle) when ^{3}H-TdR was injected between E 16 and E 22. Injection times earlier than E 16 were not investigated, however. Altman and Bayer (1978d) also observed ependymal cells in the hypothalamic third ventricle when ^{3}H-TdR was injected from E 16 on. Rakic and Sidman (1968) described labeled ependymal cells in the caudal part of the third ventricle in mice, i.e., in the immediate vicinity of the sub-commissural organ, when ^{3}H-TdR was injected starting at E 11. Finally, labeled ependymal cells were observed in the third ventricle of 25-day-old rats even after ^{3}H-TdR injection at E 10 (Korr 1978a).

On the whole, it can be concluded from these results that neuroepithelial cells which later develop into ependymal cells start to proliferate even before E 10 in some parts of the brain, probably soon after formation of the neural plate. Again, this finding is in conflict with the aforementioned concept of Fujita, i.e., the notion that the common precursor cell type proliferates in stages at different points during embryogenesis.

End of proliferation. It is difficult to determine the time at which proliferation ends, because data published on the proliferation of ependymal cells in the CNS of untreated young and adult rodents are contradictory. For instance, when Smart (1961) injected 3-day-old mice with ^{3}H-TdR, he found no labeled ependymal cells in the lateral ventricle 32 days later. Imamoto et al. (1978) administered three ^{3}H-TdR injections at 7-h intervals to young rats (19 - 25 days old) and also found no labeled ependymal cells in the lateral ventricle 1 - 35 days after the last injection. After ^{3}H-TdR injection in adult rats, none of these cells were labeled in either the third ventricle (Altman 1963) or the central canal of the spinal cord (Kerns and Hinsman 1973). In contrast to these findings, (Kulenkampff 1958; Kulenkampff and Kolb 1957) described mitoses of ependymal cells in the spinal cord of adult mice. Adrian and Walker (1962) also observed labeled ependymal cells in the spinal cord of adult mice after ^{3}H-TdR injection. Kraus-Ruppert et al. (1975) found a value of about 8% (spinal cord) and even 22% (forebrain without further topographic details) in 4-week-old mice after multiple ^{3}H-TdR injections over a 30-day period. Further quantitative data on the degree of proliferation are also available for the ependymal cells of various regions of the lateral and third ventricles in young rats (1 - 42 days old) (Chauhan and Lewis, 1979): The labeling index ranged from 2.6% (anterior lateral ventricle of 1-day-old rat) down to about 0.3% (third ventricle of 42-day-old rat). In our short-term experiments (Korr 1978a), a few labeled ependymal cells were observed in the lateral ventricle after ^{3}H-TdR injection both in 15-day-old and in 5- to 7-week-old mice (NMRI strain).

In contrast to these contradictory observations in untreated animals, unequivocal results have been obtained on the labeling of ependymal cells with ^{3}H-TdR.

After wounding. Altman (1962b) described labeled ependymal cells in large numbers in the lateral and third ventricles of 3-month-old rats after he had produced a lesion in the region of the lateral geniculate body and then injected ^{3}H-TdR into

11

the wound. The labeled ependymal cells were observed over a period from 1 to 60 days after ^{3}H-TdR injection; the intensity of the labeling decreased with time, while the number of labeled cells increased. Interestingly, Adrian and Walker (1962) found an increased number of labeled ependymal cells after stab-wounding and ^{3}H-TdR injection only when the lesion also involved the central canal. This observation could explain why Walker (1963) found labeled ependymal cells in only one animal after the production of brain lesions in adult mice. Moreover, Kerns and Hinsman (1973) reported mitoses and labeled ependymal cells after sciatic neurectomy and subsequent ^{3}H-TdR injection in young adult rats.

In summary, then, it appears that most ependymal cells and their precursors proliferate prenatally and for a short time after birth. But a few of these cells are still capable of proliferating even in the brain of adult rats and mice. One interpretation of this finding is that new ependymal cells are added because the increase in brain volume, though slight, is possibly associated with an enlargement of the ventricles (cf. also Chauhan and Lewis 1979). However, the possibility must also be considered that aged ependymal cells that are no longer functional are being replaced. The low proliferative activity observed would then be an expession of cell renewal.

3.2.5 Epithelial Cells of the Choroid Plexus

The choroid plexus consists of a highly vascular leptomeninx (i.e., mesodermal connective tissue) which is covered by a single (ectodermal) epithelial layer. The epithelial cells are derived from the ependyma and thus from the neuroepithelial cells. Because the choroid plexus is usually highly involuted, the two cell layers can seldom be differentiated by light microscopy. Thus, it is often a difficult matter to determine the labeling behavior of the individual elements (here, the epithelial cells).

Start of proliferation. Miale and Sidman (1961) observed ^{3}H-labeled epithelial cells of the choroid plexus in the fourth ventricle of fetal mice (E 17) after injecting ^{3}H-TdR at E 11. In our investigations (Korr 1978a), labeled cells of this type were found in the lateral ventricle as well as the fourth ventricle of the rat after ^{3}H-TdR injection at E 10 or later. As in the case of the ependymal cells, it follows from these findings that some precursor cells are already proliferating at a time which is close to that at which the neural plate is formed.

End of proliferation. According to Altman (1969a), the epithelial cells of the choroid plexus are capable of proliferation until about the end of the 2nd week of postnatal life. This is consistent with our own observations that labeled epithelial cells can be found in the mouse lateral and fourth ventricles in short-term experiments with ^{3}H-TdR only up to the 10th day after birth, but not in the brain of 15-day-old or adult mice (Korr 1978a). The results of Chauhan and Lewis (1979) suggest that the end of proliferation occurs later in the course of ontogeny: These authors reported labeled plexus epithelial cells fo the lateral ventricle up to age 42 days, though older animals were not investigated. The labeling index declined from 0.63% (1-day-old rat) to 0.07% (42-day-old rat). As Altman (1963) showed, however, labeled epithelial cells of the choroid plexus could not be found even 2 weeks after ^{3}H-TdR injection in the brain of adult, 4-month-old rats.

The findings quoted above are in complete disagreement with the results of Mares (Mares and Lodin 1974; Mares et al. 1975): In short-term experiments with ^{3}H-TdR, labeled epithelial cells as well as labeled endothelial cells of the choroid plexus were found in the lateral ventricle of mice between 1 and 12 months old.

Despite these latter findings, however, it is probably correct to assume that the epithelial cells of the choroid plexus cease proliferating during the first weeks of life, possibly at the onset of adolescence.

We should also note the results obtained for another cell type derived from the ependyma: the pineal cells. According to studies by Dill and Walker (1966), the labeling index of these cells is between 25% and 50% in 2-day-old mice; an index of about 2% was measured in 9-day-old mice, and less than 1% in 21-day-old animals. If the ^{3}H-TdR was injected into the adult animal, labeled pineal cells could no longer be observed. A decline in the number of labeled pineal cells with age was also reported by Wallace et al. (1969) in the rat brain. Nevertheless, a few labeled pineal cells were still found by these authors in 120-day-old rats after injection of ^{3}H-TdR.

3.2.6 Endothelial Cells

Start of proliferation. Our investigations point to a rather early start of endothelial cell proliferation, before about E 11: After ^{3}H-TdR injection at E 11, a few weakly labeled endothelial cells were found in certain brain areas of the 25-day-old rat (e.g., the cortex or n. ruber) (Korr 1978a). Heavily labeled endothelial cells were not observed in this study until ^{3}H-TdR was injected at E 15. Brückner et al. (1976) reported the presence of heavily labeled "endothelial-like cells" both in the dorsal and ventral nucleus of the lateral geniculate body and in the superior colliculus of young rats (P 22 to P 24) when ^{3}H-TdR had been injected at E 12 or later.

In connection with the autoradiographic data presented, it is of interest to note that the pial blood vessels begin to penetrate the neural wall at this early stage of embryonic development (around E 11 or 12) (Bär and Wolff 1972).

End of proliferation. Generally speaking, some labeled endothelial cells can always be seen in the brain of untreated adult rats and mice in short-term experiments with ^{3}H-TdR (Altman 1963; Mares and Lodin 1974; Korr et al. 1973, 1975). A few labeled endothelial cells have even been found in the brain of senile mice, as in the case of the neuroglia (Korr 1978a). This shows that the endothelial cells are also capable of proliferating throughout the lifetime of the animal.

3.2.7 Microglia and Pericytes

According to the classic concept of Rio-Hortega (1921, 1932), the microglia migrate into the brain toward the end of fetal development. The question of whether the microglia are of mesodermal or ectodermal origin will not be discussed here (for this, see Vaughn and Peters 1971; Kitamura et al. 1972; Kitamura 1973; Matthews 1974; Sturrock 1974c; Privat 1975; Oehmichen 1975; also see the recent experimental findings of Blakemore 1975; Fujita and Kitamura 1975; Hager 1975; Oehmichen et al. 1975; Torvik 1975; Kitamura et al. 1977; Imamoto and Leblond 1977, 1978; Adrian

13

et al. 1978; Sturrock 1978a,c; Boya et al. 1979). We shall also exclude the question of whether the pericytes should be considered a form of microglia ("pericytal microglia": Mori and Leblond 1969a; cf. also Hager 1968; Jones 1970; Mori 1972; Brichova 1972; Baron and Gallego 1972; Sturrock 1974c; Boya 1976) or an independent cell form (Vaughan and Peters 1974; Stensaas 1975).

Start of proliferation. Few autoradiographic studies have been done on the prenatal origin of microglia and pericytes: As mentioned in Sect. 3.2.2, Berry and Rogers (1966) observed labeled microglial cells in the cortex of young rats after ^{3}H-TdR injection at E 18. However, in our experiments with two injections of ^{14}C-TdR at E 19 and E 20, we could not find in the forebrain of the 30-day-old offspring any microglial cells or pericytes which could clearly be scored as labeled. The observation of many cells with one or two ^{14}C-tracks could indicate that these cells were indeed proliferating at the time of the ^{14}C-TdR injections, but their labeling was so diluted by many successive divisions that it fell below the counting limit of three tracks per nucleus. This limit was derived from the observation that at most two tracks per nucleus occur as nonspecific background over unlabeled areas (Korr, to be published).

Sturrock (1974c) showed by electron microscopy (without autoradiography) that pericytal microglial cells could be found as early as E 19 in the mouse anterior commissure. The first interstitial microglial cells were observed here at birth.

In contrast to the aforementioned concept of Rio-Hortega 1921, 1932), von Sántha (von Sántha 1932; von Sántha and Juba 1933) also demonstrated microglial cells in the brain of embryonic rabbits and rats with the aid of metallic impregnation reported by Rio-Hortega 1921 (in rats, microglial cells first appeared in embryos of about 15 mg body weight). Thus, these microglial cells seemed to appear at about the same time as the vascularization of the neural wall began. Kershman (1939) made a similar observation in human embryos. In a recent electron microscopic study Sturrock (1978c) found microglial cells in the brain of the mouse embryo as early as E 13. As the author suggests these microglial cells — similar to the microglial cells observed at later periods of fetal development — might be derived from intraventricular macrophages which for their part might possibly be formed within the choroid plexus.

End of proliferation. With the aid of a combined method (Rio-Hortega method of silver carbonate impregnation followed by autoradiography, see Korr 1978b), it could be shown that about 50% of all microglial cells as well as pericytes in the forebrain of 45-day-old mice were labeled after repeated injections of ^{14}C-TdR between P 10 and P 14. On the other hand, if the young mice were injected with ^{14}C-TdR between P 15 and P19, only about 5% of the microglia and pericytes were labeled in the brain of the 45-day-old animals. However, no microglial cells or pericytes could be labeled by multiple ^{14}C-TdR injections between P 20 and P 24 or between P 41 and P 45 (Korr, to be published). These cell types were also unlabeled in 5- to 7-week-old mice 1 h as well as 14 days after ^{14}C-TdR injection (Korr 1978a). Thus, it is apparent that in the brain of the untreated mouse microglial cells and pericytes or their precursors have completed their proliferation phase about 3 weeks after birth.

This conclusion is further supported by the following results: Imamoto and Leblond (1978) studied over a 35-day period the corpus callosum of young rats which had received ^{3}H-TdR three times at the age of 5 days. Over the whole period studied (excluding 2 h after the last ^{3}H-TdR injection), they always found a high percentage of labeled microglial cells and pericytes. These authors also showed that the observed labeled microglial cells originated from monocytes, which had been transformed first into macrophages (ameboid cells) and then into microglial cells. Similar studies, i.e., injection of ^{3}H-TdR in 19- to 20-day-old rats, led to a significantly smaller percentage of labeled microglial cells and pericytes (Imamoto et al. 1978). Finally, these investigations showed that no labeled microglial cells or pericytes were found later (with very few exceptions) in rats which were not injected until the age of 25 days. Microglial labeling was also absent in the corpus callosum of adolescent rats (60-90 g body weight 2 h and 28 days after ^{3}H-TdR injection in light microscopic autoradiographic studies of semithin sections (Imamoto and Paterson 1974) as well as in electron microscopic autoradiographic studies (Mori and Leblond 1969a).

Compared with the aforementioned results of Leblond and co-workers, it is remarkable that Paterson and Leblond (1977) observed a large number of labeled microglial cells and pericytes (total of about one-quarter of all these cells) in the supraoptic nucleus of young rats which had received the first ^{3}H-TdR injection at age 30-35 days and then two more injections daily over a period of 14 days. This finding is referred to two animals which served as controls in special drinking experiments with hypertonic saline solutions. However, no labeled microglial cells or pericytes were found in the caudate nucleus of these animals.

Our own experiments cited above (Korr, to be published) conflict with the results of Hommes and Leblond (1967): These authors described labeled microglial cells in the brain of *adult* rats 9 h after injection of ^{3}H-TdR. However, the authors themselves point out that these cells were only tentatively identified. There is also the view of Cammermeyer (1970), who emphasizes in his survey work that microglia and pericytes are capable of proliferating in the brain of untreated adult rodents. He substantiates this claim with studies on paraffin sections. However, the different cell types are differentiated solely on the basis of criteria of nuclear morphology, a method which is no longer considered reliable (cf. Mori and Leblond 1969a,b, 1970; Ling et al. 1973).

The proliferation of microglia and pericytes in the adult animal brain might well be possible subsequent to *wounding* of the brain. It must be remembered in this connection, however, that the identification of cell types is an uncertain matter even in modern electron microscopic studies; it is particularly difficult to distinguish between microglial cells ("reactive microglia") and immigrated hematogenous cells such as macrophages and monocytes (cf. Peters et al. 1970; Blakemore 1972; Kitamura et al. 1972; Kitamura 1973; Adrian and Williams 1973). Thus, labeled microglial cells and pericytes found after a brain lesion may stem from immigrated hematogenous monocytes which have been transformed into microglialike cells. On the other hand, Oehmichen et al. (1973) and Kitamura et al. (1978) have clearly shown that (in the brain of adult rabbits after stab wounding) local or "resting" microglial cells are also capable of proliferation.

Based on all the available evidence, it appears correct to conclude that the proliferation phase of microglia and pericytes in the brain of untreated rats and mice ceases with the onset of adolescence, i.e., at about 3 weeks after birth in mice. As for results which would indicate that these cells also proliferate in untreated adult animals, one must consider the possibility that the animals were not healthy at the time of the study (viral infection?).

3.2.8 Meningeal Cells

The term "meningeal cells" encompasses various cell types within the leptomeninx: cells of the pia mater, cells of the subarachnoid space, cells of the arachnoid, and finally cells belonging to the blood vessel system (endothelial cells, adventitial cells, or pericytes). Immigrated hematogenous cells are also often found in the region of the leptomeninx.

Start of proliferation. Only a few studies have been done on the proliferation of the cells at the boundary between the brain and cranial bone. In the view of Altman (1969a), most of these cells are produced postnatally in the rat, approximately before the end of the 1st month of life. The youngest animals which Altman studied were neonate rats. When he injected animals of this age with ^{3}H-TdR, he later found labeled meningeal cells in the adult animal. Meningeal cells of neonate mice can also be labeled in short-term experiments (Korr 1978a).

No data are yet available in this area for the prenatal phase of life. It may be supposed, however, that some of these cells also proliferate prenatally, especially toward the end of fetal development.

End of proliferation. No precise data are available. Some labeled meningeal cells are demonstrable 1 h after injection of ^{3}H-TdR in 5- to 7-week-old mice (Korr 1978a). Oehmichen and Grüninger (1974) determined the labeling index of arachnoidal, trabecular, adventitial, and pial cells in 6-week-old rabbits. They obtained similar values for all cell types investigated (labeling index ca. 0.2%).

3.2.9 Graphic Summary of the Proliferation of the Different Cell Types

Disregarding the meningeal cells (for which data are still too sketchy), the results of the preceding sections are summarized in graphic form in Fig. 21; they will be described in more detail later. The approximate degree of proliferation is also shown. On the basis of the scheme in Fig. 21, the different cell types in the brain of untreated rats and mice can be divided into two main groups:

1. The precursor cells of neurons, the ependymal cells, the epithelial cells of the choroid plexus, and the microglial cells and pericytes proliferate mainly prenatally, but also postnatally for a brief period of a few weeks (e.g., during the first 3 weeks after birth in mice) until the start of adolescence. Except for a few ependymal cells, these cell types no longer proliferate in adulthood.

2. The astrocytes, oligodendrocytes, cells of the subependymal layer, and the endothelial cells are capable of proliferation throughout postnatal life. Some cells of these types are always proliferating, even in the brain of very old animals, though the number of cells proliferating at any one time is very small.

4 Cell Cycle Parameters of the Different Cell Types in the Rodent Brain During Pre- and Postnatal Ontogeny

4.1 Methods for the Experimental Determination of Cell Cycle Parameters

Of the methods available for determining the duration of the cell cycle and its phases (survey: e.g., Nachtwey and Cameron 1968; Cameron 1971; Aherne et al. 1977), our discussion will center on those which are suitable for the study of proliferative processes in the brain (cf. Sidman 1970).

4.1.1 Percent Labeled Mitoses Method

The duration of the cell cycle and its phases is in most cases determined by the percent labeled mitoses method (PLM method; Quastler and Sherman 1959). In this method the cell cycle parameters are derived from the percentage of labeled mitoses as a function of time after a single injection of ^{3}H-TdR. Details on the determination of the individual cell cycle phases are given by Steel (1977), among others.

This method has been employed in studies on brain cell proliferation in cases where sufficient mitoses can be found (e.g., the neuroepithelial cells during embryogenesis). However, this method is unsuitable for brain cells of adult rodents outside the subependymal layer, first because there are very few mitotic figures, and second because their detection depends too much on the quality of the fixation (cf. Fleischhauer 1967; Cavanagh and Lewis 1969).

4.1.2 Continuous Infusion with ^{3}H-Thymidine

Fujita (1962, 1963, 1966) showed in embryonic neuronal tissue from chickens and mice that cell cycle parameters can be deduced from the increase in the percentage of labeled cells when ^{3}H-TdR is administered by continuous infusion over a period which corresponds at least to the difference between the cell cycle time and the length of the S phase. It is particularly important in this technique to determine the time at which all cells present are labeled.

However, this method is unsuitable for determining cell cycle parameters of the different cell types in the brain of young and adult rodents, for it is impossible to label all cells of a particular histologic type by continuous labeling with ^{3}H- or ^{14}C-TdR over prolonged periods of time (cf. Kraus-Ruppert et al. 1973).

Thus, a common feature of the two methods cited in Sects.4.1.1 and 4.1.2 is that they are generally unsuitable for studies in the adult animal brain. We will therefore turn our attention to techniques which can also be employed in the aforementioned cases. Because these methods have not commonly been used in brain cell research or are new, they will be described in more detail.

17

4.1.3 Grain-Count Halving Method

The mean cycle time (T_C), especially for blood cells, is sometimes determined from the decline in the mean number of grains per nucleus as a function of time following a single injection (mainly) of ^{3}H-TdR (Alpen and Cranmore 1959; Killmann et al. 1962; Baserga et al. 1963; Fried 1968, 1970). This method is based on the principle that the labeled DNA of the mother cell is distributed more or less uniformly between the two daughter cells. As a result, the grain count per nucleus is reduced roughly to half its value after each complete division. For the problems involved in the relationship between the grain count per nucleus and the quantity of labeled DNA see Schultze (1969, pp 89 and 161). It is important to note that this method does not rely upon the finding of mitoses.

As was shown for neuroglial cells, endothelial cells, and the cells of the sub-ependymal layer in the brain of adult mice (Korr et al. 1975; Korr 1978a), this grain-count halving method (Fried 1968) is entirely suitable for brain cell studies. It will be shown by way of examples how the T_C can be determined from mean grain counts per nucleus measured at various times after injection of ^{3}H-TdR. "Mean grain count" generally refers to the arithmetic mean, although other values such as the median grain count or geometric mean can also be used (cf. Fried 1970). It should also be noted that a precise determination of T_C requires that an assumption must be made regarding the mode of proliferation of the investigated cells; i.e., it must be known whether the growth of these cells is exponential or a steady-state.

1. In the case of *exponential growth* of the labeled cells, each labeled mother cell gives rise to two labeled daughter cells, each of which proliferates further. Thus, the labeled cells undergo continuous division. After an injection of ^{3}H-TdR, all cells are labeled which are in the S phase at the time of injection. The mean grain count per nucleus remains constant over a period corresponding to the duration of the phases G_2 + M (T_{G_2+M}), until the first labeled cells have undergone mitosis. When all the labeled cells have divided after a time T_{G_2+M+S}, the mean grain count per nucleus has been reduced overall by a factor of 2. During the subsequent interval $T_{G_2+M+G_1}$, the grain count per nucleus again assumes a constant value. After the labeled cells have undergone the second mitosis after labeling, the mean grain count per nucleus is again reduced by a factor of 2, and so on. As Eq. (1) shows, the overall decline in the mean grain count per nucleus is exponential:

$$A_n = A_0 \cdot \frac{1}{2^n},$$
(1)

where A_n is the mean grain count per nucleus after n = 1, 2, 3, ... mitotic divisions, and A_0 is the mean grain count per nucleus shortly after ^{3}H-TdR injection.

Figure 1 shows how this method is applied to cells undergoing exponential growth in the subependymal layer of the lateral ventricle in adult mice (see Korr 1978a). Here the data points (each corresponding to the value measured for one animal) are plotted on a semilog scale as a function of time after ^{3}H-TdR injection. In this case the period of time in which the mean grain count per nucleus is

18

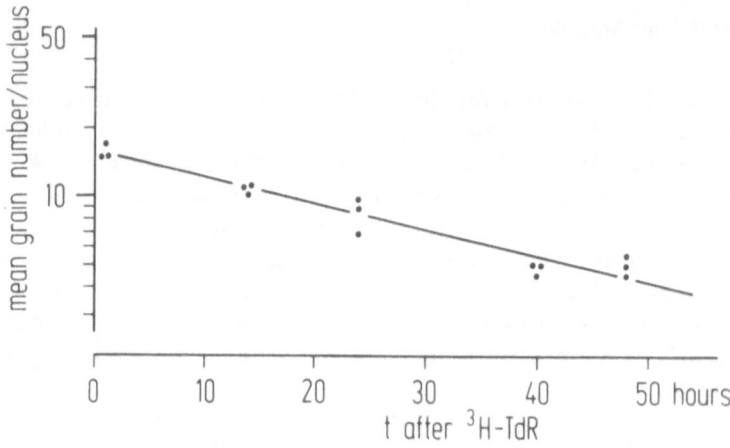

Fig. 1. Mean grain count per nucleus for labeled subependymal cells in the brain of the adult mouse as a function of time after ^{3}H-TdR injection. Each *point* corresponds to the mean grain count per nucleus for one animal

reduced to half its value (grain-count halving time) can be determined from the straight line drawn through the data points. Of course this period can be considered only a rough estimate of T_C: By drawing a straight line through the data points, the assumption is made that the labeled cells are distributed uniformly over the cell cycle shortly after ^{3}H-TdR injection. Owing to the variations that always exist in the length of the cell cycle phases, this will be approximately the case as soon as the labeled cells have undergone several cycles, but certainly not in the first cycle after labeling.

From the graph in Fig. 1, a mean cycle time of about 25 h was determined. As will be shown, this value is too high. This brings us to another point which must be considered when determining T_C from the decline in the mean grain count per nucleus: Autoradiographs made at different times during an experiment can be compared with each other only if they were produced under identical conditions (i.e., identical conditions of exposure, development and fixation). Only when this requirement is met can mean grain counts be compared with the aid of correction factors for different exposure times or different ^{3}H-TdR dosages. In addition, care must be taken that the autoradiographs are evaluated under similar conditions: The frequency distributions of the grain count per nucleus should largely coincide for the experiments to be compared. This is of primary importance in the range of small grain counts. Here it must be considered that a portion of the weakly labeled cells may not be recorded; the size of this component can vary considerably in different experiments. But the range of high grain counts should also be in agreement. Inaccuracies in grain count determinations are more likely to occur in this range than at lower grain densities. Grain density saturation effects must also be allowed for.

These considerations are illustrated by the following example: Figure 2 shows in histogram form the relative frequencey distributions of the grain count per nucleus for labeled cells of the subependymal layer in the adult mouse brain after injection of ^{3}H-TdR. The mean grain counts per nucleus in these experiments are

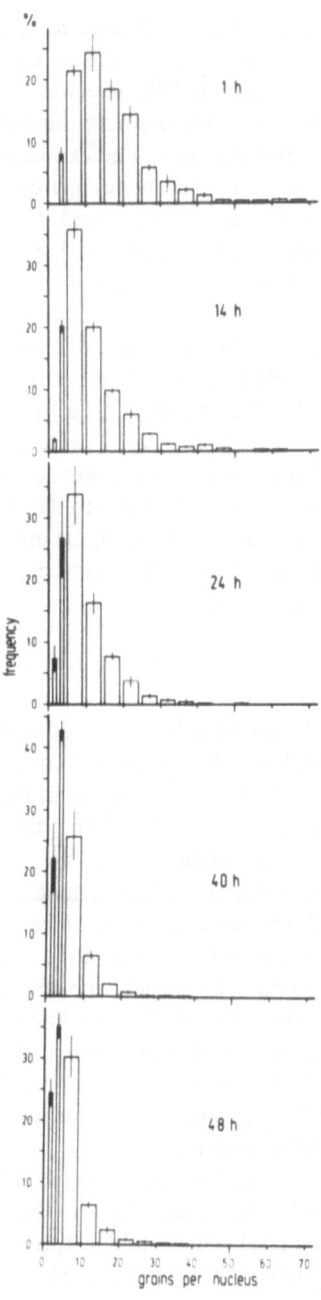

Fig. 2. Relative frequency distributions of grain count per nucleus for labeled cells of the subependymal layer in the brain of the adult mouse after injection of 500 μCi ^{3}H-TdR. Exposure time of autoradiographs: 4 days. The mean value of three individual distributions is given for each grain-count interval. The *vertical lines* indicate the standard error of the mean value

given in Fig. 1. As Fig. 2 shows, the peak of the histograms shifts with time toward the range of lower grain counts. While the portion of very weakly labeled cells (i.e., cells with 1-2 grains per nucleus) is still 0% 1 after ^{3}H-TdR injection, it increases to about one–quarter of all cells 48 h after injection. These counts suggest that all the labeled cells were no longer detected 48 h after injection. The exposure time of the autoradiographs was apparently insufficient to reveal subependymal cells that were weakly labeled. Thus, the mean grain count per nucleus measured 48 h after injection was too high. As a result, the line shown in Fig. 1 should have a greater slope. This means that the actual cycle time of the cells of the subependymal layer in the adult mouse brain is less than 25h.

2. In the case of *steady state growth,* only one of the two daughter cells produced by the mitosis proliferates further, while the other daughter cell is non-proliferating but still constitutes a part of the observed cell population. The result of this is that the mean grain count per nucleus of the *proliferating* cells decreases with time in the same manner described in example 1. But when the mean grain count of *all* labeled cells is determined, the labeling of those cells which do not proliferate further after each mitotic division (i.e., do not further reduce their grain count) must also be taken into account. On the whole, therefore, the mean grain count per nucleus of all labeled cells $(A_n *)$ decreases after further mitoses according to Eq. (2):

$$A_n * = A_0 \cdot \frac{1}{n+1} \quad \text{[symbols as in Eq. (1)].} \tag{2}$$

This means that the mean grain count per nucleus of all labeled cells is still reduced to half its value after the first mitosis following labeling, but is only reduced to one–third of the original value after the second mitosis instead of one–quarter as in the case of exponential growth, and so on. Hence a straight line through the data points plotted on a semilog scale would give a value for T_C which is too high.

Figure 3 shows how an acceptable estimate of T_C can nevertheless be obtained for the case of steady state growth, taking the glial cells in the cortex and corpus callosum of 14-day-old rats as an example (cf. Korr et al. to be published b). To do this, it is necessary to obtain further information on the proliferating cells, namely the length of the S phase (T_S) and G_2+M phase (T_{G_2+M}). These values can then be used to construct theoretical curves for the decline of the mean grain count per nucleus, where T_C is the only parameter that can be freely chosen. The object is to find the theoretical curve (constructed with the values measured for T_S and T_{G_2+M} and a certain assumed value for T_C) which best agrees with the values determined experimentally.

The symbols in Fig. 3 correspond to the measured mean grain counts per glial cell nucleus in the cortex (dots) and corpus callosum (triangles) of 14-day-old rats. These values agree most closely with a theoretical curve (broken line) which was constructed under the following assumptions confirmed experimentally beforehand (except for T_C):

a) When labeled cells undergo one mitosis, their mean grain count per nucleus is reduced to half its value.

b) On the average only one of the two daughter cells resulting from a mitosis proliferates further. The daughter cell which does not proliferate further remains a part of the cell population, however.

c) The cells proliferate with the following cell cycle parameters: T_S = 10 h, T_{G_2+M} = 4 h, T_C = 20 h. It was also assumed that no variations occurred in the

21

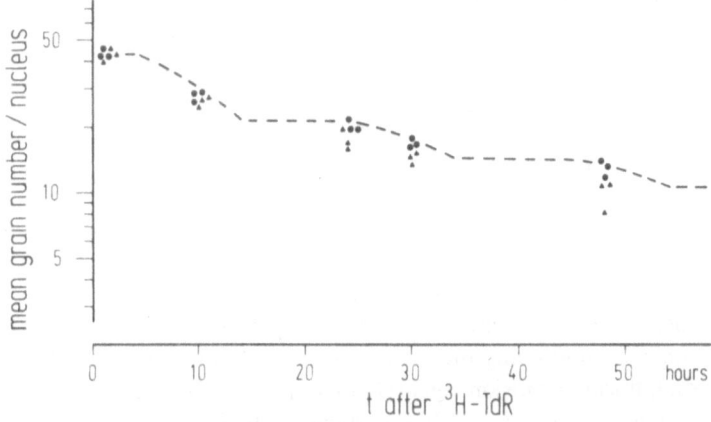

Fig. 3. Mean grain count per nucleus for glial cells in the cortex (•) and corpus callosum (▲) of the 14-day-old rat as a function of time after ^{3}H-TdR injection. *Dots and triangles,* experimental values for individual animals. *Broken line,* theoretical curve (see text)

length of the individual cell cycle phases. This assumption was made only to simplify the construction of the curve. The presence of cycle variations which are quantitatively unknown but certainly present has the effect of changing the step-by-step decline of the mean grain count into a continuous decline.

If the experimental values in Fig. 3 are compared with theoretical curves constructed under the assumption of $T_C = 15$ h or $T_C = 25$ h, large discrepancies arise. This shows that the glial cells in the investigated brain regions of 14-day-old rats proliferate with T_C of approximately 20 h.

4.1.4 Double-Labeling Methods with ^{3}H- and ^{14}C-TdR

Before discussing double-labeling methods with ^{3}H- and ^{14}C-TdR for the determination of T_s and T_c, we shall first describe the basic technique for evaluating such experiments: namely the two emulsion layer autoradiography.

4.1.4.1 Two Emulsion Layer Autoradiography

Two emulsion layer autoradiography is a technique used to discriminate between ^{3}H- and ^{14}C-labels by virtue of the different ranges of their β-particles (^{3}H-β-particles with the maximum energy have a range of 0.54 mg/cm^2 in tissue with a density of 1, and ^{14}C-β-particles a range of 29 mg/cm^2). The β-particles of the two isotopes are demonstrated with sufficient accuracy in one of the two emulsion layers when the ^{3}H:^{14}C activity ratio is sufficient large (e.g., about 50 times more ^{3}H activity than ^{14}C activity; for details see Schultze et al. 1976).

The basic structure of the two emulsion layer autoradiograph is shown schematically in Fig. 4. The labeled section is first coated with a thin emulsion layer by dipping (about 1-2 μm thick in a dry state before developing). This thin emulsion layer serves to record the short-range β-particles emitted by the ^{3}H. The first emulsion layer is developed after an exposure time of several days. The ^{3}H:^{14}C activity ratio must be sufficient to ensure that enough silver grains are present after this brief period

22

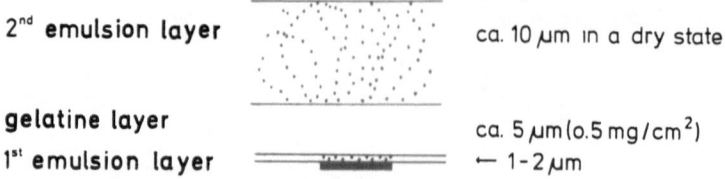

2nd emulsion layer .... ca. 10 μm in a dry state

gelatine layer ca. 5 μm (0.5 mg/cm^2)
1st emulsion layer ← 1-2 μm

Fig. 4. Schematic diagram of the two emulsion layer autoradiograph

to permit evaluation. Then the first emulsion layer is coated with a gelatin layer about 5 μm thick. The purpose of this gelatin layer is to prevent all ^{3}H-β-particles, even those with the maximum range, from entering the second emulsion layer. This gelatin layer is covered with a second, thicker emulsion layer (about 10 μm in a dry state before developing). This layer produces a record of ^{14}C-β-particles in the form of ^{14}C-β-tracks. This second emulsion layer can be exposed for a sufficiently long period (several weeks or months) to ensure that enough tracks are present for evaluation, since the thin emulsion layer is already developed and no additional silver grains can appear there.

Figure 5 shows examples of two emulsion layer autoradiographs with the three types of labeling. The first emulsion layer is in focus in the pictures on the left, while the second, thick emulsion layer is in focus in the pictures on the right. Silver grains in both layers (seen as silver grain tracks in the second emulsion layer) characterize a cell which is double labeled with both ^{3}H and ^{14}C (top). If silver grains are present only in the first emulsion layer, with no tracks in the second layer, the cell is purely ^{3}H-labeled (center). Finally, tracks in the second emulsion layer, with few or no silver grains in the first layer, signify that the cell is purely ^{14}C-labeled (bottom).

Figure 5 further shows that in the case of ^{14}C-labeling, be it a double or pure ^{14}C-label, the ^{14}C-β-tracks extend far beyond the area of the nucleus. This explains why an accurate recording is impossible if the ^{14}C-labeled cells are too close together. For this reason double-labeling experiments involving the quantitative evaluation of cells with different labels can generally be done only on tissues in which the individual labeled cells are located in a sufficient distance from one another. Thus, double-labeling experiments are unsuitable for cell types in the embryonic brain, as well as for cells such as those in the subependymal layer of lateral ventricle of adult rats and mice. The double-labeling method should be reserved for studies in which labeled cells are sparse, as in the case of the neuroglia and the endothelial cells in the brain of adult rodents.

4.1.4.2 Determining the Length of the S Phase

With the aid of Fig. 6, it will be shown how the length of the S phase (T_S) can be determined by double labeling with ^{3}H and ^{14}C. This method was first employed independently by Hilscher and Maurer (1962) as well as Pilgrim and Maurer (1962) and Wimber and Quastler (1963). It is important to note that this determination of T_S does not depend upon the presence of mitoses.

In this double-labeling method the animals receive a first injection of ^{14}C-TdR. All cells in the DNA synthesis phase at the time of injection become ^{14}C-labeled. After an interval Δt of a few hours duration (e.g., 2 h in Fig. 6), the animals receive a second injection, this time of ^{3}H-TdR. This labels all cells with ^{3}H which were in the S phase at that time. As a result of these two injections, three groups of cells are obtained, each with a different type of label:

23

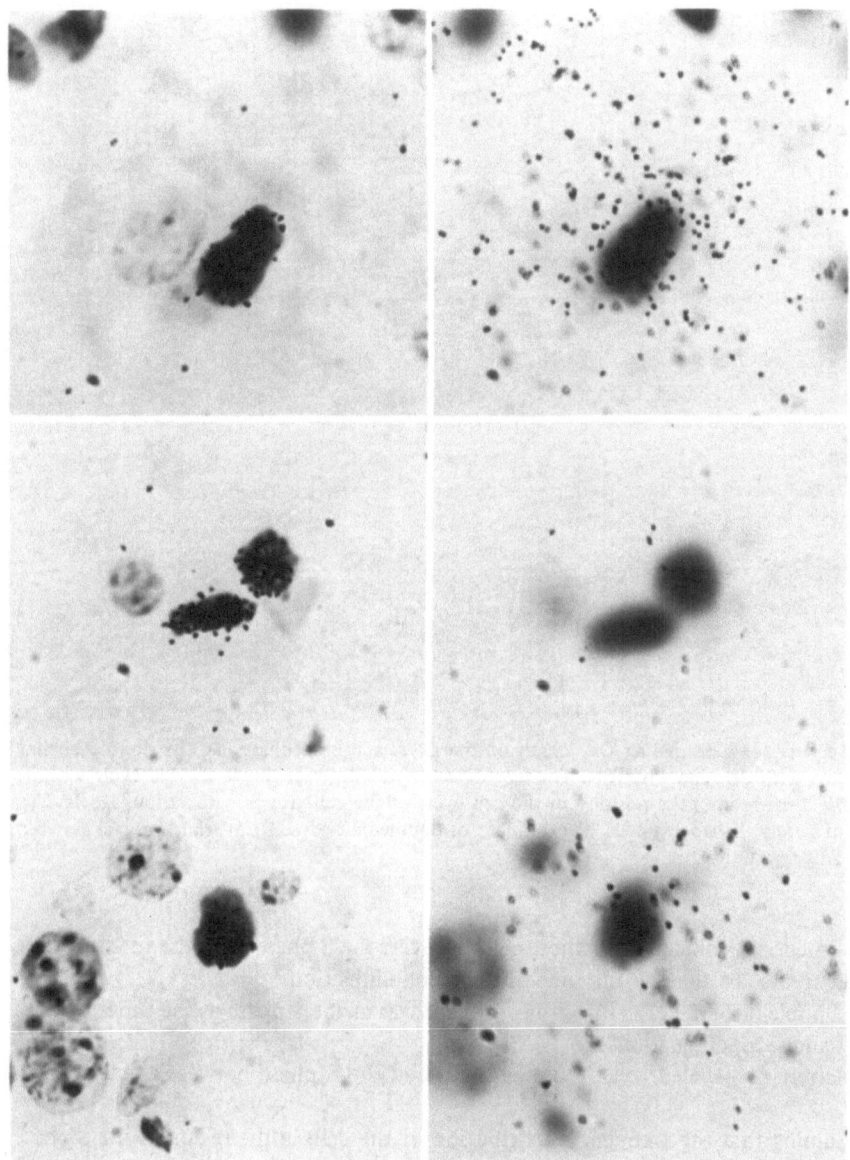

Fig. 5. Labeling of glial cells in the two emulsion layer autoradiograph. Each picture on the left shows the first emulsion layer in focus, while the second (thick) emulsion layer is in focus on the right. *Top,* double-labeled glial cell; *center,* purely ^{3}H-labeled pair of glial cells; *bottom,* purely ^{14}C-labeled glial cell. All neuroglial cells shown originate from the cortex of the adult mouse. The animal received an initial injection of 800 μCi ^{3}H-TdR, 10 μCi ^{14}C-TdR 4 h later, another 5 μCi ^{14}C-TdR 15 h after the start of the experiment, and was killed 1 h after the last injection. Feulgen stain. Exposure time for first emulsion layer: 3 days, for second emulsion layer: 18 days. X 1300

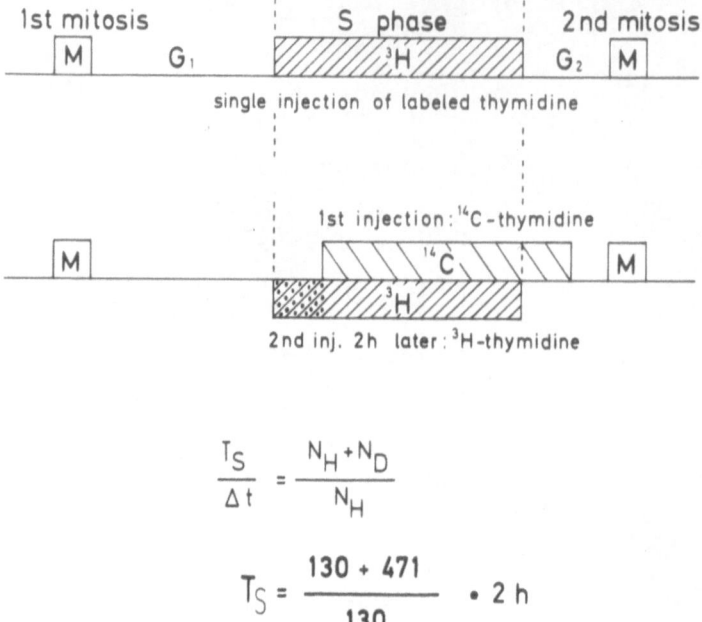

$$\frac{T_S}{\Delta t} = \frac{N_H + N_D}{N_H}$$

$$T_S = \frac{130 + 471}{130} \cdot 2\,h$$

$$\underline{T_S = \quad 9.3\ \text{hours}} \quad \text{glial cells , mouse}$$

Fig. 6. Scheme for determining the length of the DNA synthesis phase (T_S) by double-labeling experiments with [3]H- and ^{14}C-TdR. The *upper part* of the diagram shows the cell cycle with its phases; *the center* shows the position in the cell cycle of the cell groups with various labels. N_H, number of purely ^{3}H-labeled cells; N_D, number of double-labeled cells; Δt, time interval between the two TdR injections

1. Purely ^{14}C-labeled cells: these cells have left the S phase after the first injection and were in the G_2-phase at the time of the second injection.

2. Double-labeled cells: this group of cells was in the S phase at the timer of both the first and second injections.

3. Purely ^{3}H-labeled cells: these cells entered the S phase between the two injections.

Assuming that the frequency distribution of the cells in the region of the S phase is constant (as is true in steady state growth and approximately true in exponential growth with cycle times of 1 day or more), the number of purely ^{3}H-labeled or purely ^{14}C-labeled cells is proportional to the time difference Δt between the two injections. The number of all ^{3}H-labeled cells with and without [14]C (or all ^{14}C-labeled cells with and without ^{3}H) is proportional to T_S. As the proportional equation in Fig. 6 shows, T_S can easily be calculated from the experimental values indicated. Only double- and purely ^{3}H-labeled cells were taken into account in this equation. In the example given in Fig. 6 (neuroglial cells in the adult mouse brain), there were 130 purely ^{3}H-labeled and 471 double-labeled glial cells, with a time difference $\Delta t = 2$ h; thus, $T_S = 9.3$ h (cf. Korr 1973; Korr et al. 1973).

As stated earlier, the grain-count halving method provides only a rough estimate of T_C. To make it possible to determine experimentally the cycle time of neuroglia and endothelial cells in the adult mouse brain in another, independent way, a special double-labeling technique with ^{3}H- and ^{14}C-TdR was developed (Korr 1974; Korr et al. 1975). This method can be called the "method of labeled S-phases" (cf. Pilgrim et al. 1966) in analogy with the "method of labeled mitoses" (Quastler and Sherman 1959, see also Sect. 4.1.1). As in the case of the double-labeling method previously described for determining T_S, this method is also independent of the presence of mitoses. Essentially, it involves the determination of the time interval between two successive S phases. This is accomplished by visualizing the passage of a small group of purely ^{3}H-labeled cells through the next S phase by administering a second label, ^{14}C-TdR.

Figure 7 shows procedural details on the method of labeled S phases. First a small group of purely ^{3}H-labeled cells is produced by injecting the animals with ^{3}H-TdR. Then a second injection, this time ^{14}C-TdR, is given after an interval of, say, 4 h. This leads to a 4-hour-wide group of purely ^{3}H-labeled cells (see part I in Fig. 7).

It is also possible, of course, to follow the passage of the entire ^{3}H-labeled S phase population through the subsequent S phase. This technique was originally described by Pilgrim et al. (1966). With an S phase length of about half the cycle time T_C, however, observations would become difficult: The first ^{3}H-labeled cells would already have reached the next S phase while those cells which were at the start of S at the time of ^{3}H-TdR injection are still in this S phase.

To observe the passage of the purely ^{3}H-labeled cells through the next S phase, the experimental animals are injected with ^{14}C-TdR at various times after the start of the experiment. Each of the animals is killed 1 h after the last ^{14}C-TdR injection. Only in those cases where the group of purely ^{3}H-labeled cells is in an S phase will the injection of ^{14}C-TdR produce additional labeling in these cells, thereby causing them to become doubly labeled. Thus, the disappearance of the group of purely ^{3}H-labeled cells and their reappearance some time later shows that these cells have passed through an S phase.

The passage of a group of purely ^{3}H-labeled cells through the S phase could be readily demonstrated in the simple manner described above if the proliferation of these cells is exponential, i. e., if *all* purely ^{3}H-labeled cells continue to proliferate after a mitotic division. In the case of steady state growth, however, a purely qualitative e-valuation would encounter difficulties: Here only half of the purely ^{3}H-labeled cells proliferate further on the average, and only these cells can be double labeled. The other half, meanwhile, cease proliferation and thus retain their pure ^{3}H-label. A qualitative evaluation would also meet with the problem that the number of proliferating cells per brain section varies greatly in different animals (in adult mice, for example, the number of labeled glial cells per forebrain section varies by more than a factor of 2; see Fig. 13 in Sect. 5.2.1.1a). Thus, the size of this cell population in different animals cannot be reliably predicted.

To avoid errors coming from these difficulties, a relative measure was selected for the number of purely ^{3}H-labeled cells present at any given time:

$$v = \frac{\text{number of purely } ^3\text{H-labeled cells}}{\text{number of double-labeled cells}}$$

This ratio v is independent of animal-specific variations in the number of labeled cells present in the section; it varies only when the cell fluxes within the cell cycle are different in different animals.

Taking the glial cells in the brain of the 14-day-old rat as an example, we shall explain the individual steps which are necessary to determine the mean cycle time (T_C) from the v ratios measured at different times after injection of ^{3}H-TdR (see part II in Fig. 7). The procedure is similar in principle to that used in determining T_C from the decline in the mean grain count per nucleus as a function of time after ^{3}H-TdR injection (see Sect. 4.1.4.2): Again, the object is to find that theoretical curve which best agrees with the measured values of v. The theoretical curve, i.e., the variation of v as a function of time after ^{3}H-TdR injection, is based on various assumptions, of which only one, T_C, can be freely chosen.

The following assumptions were made in constructing the theoretical curve in Fig. 7:

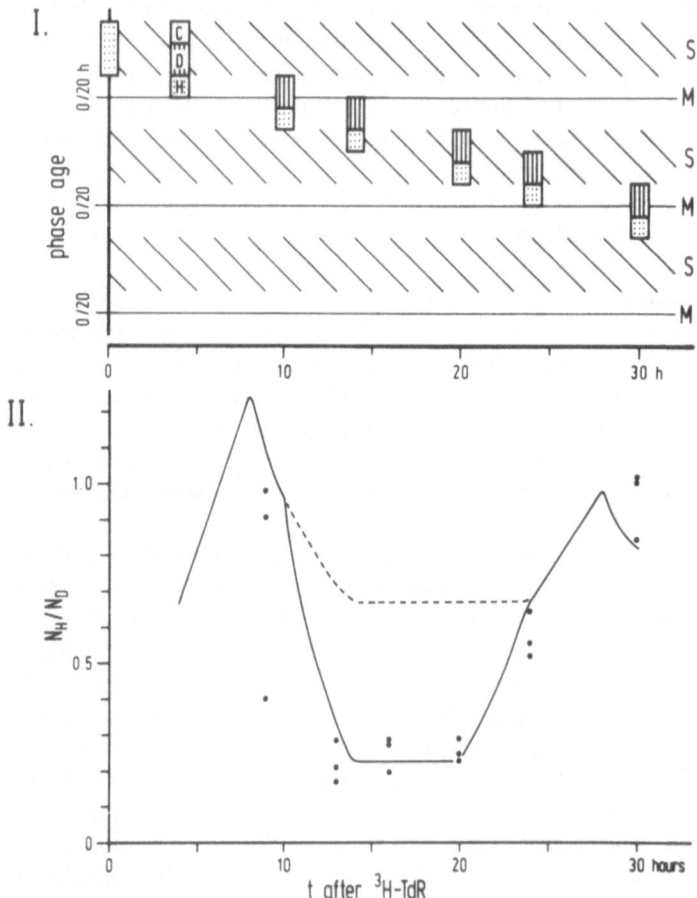

Fig. 7. Scheme and results for the determination of mean cycle time by the "method of labeled S-phases" for glial cells in the forebrain of the 14-day-old rat. I, scheme of method (see text); II, ratio of purely ^{3}H-labeled to double-labeled glial cells as a function of time after ^{3}H-TdR injection. *Dots,* experimental values for individual animals; *solid line,* theoretical curve with second ^{14}C-TdR injection; *broken line,* theoretical curve without second ^{14}C-TdR injection (for further details on curves, see text)

27

Mode of proliferation. In the case of the glial cells, only one of the two labeled daughter cells proliferates further after mitosis. For simplicity, the fact that glial cells proliferate with a certain cell loss (see Sect. 5.1.1.2) is disregarded.

Cell cycle parameters. The measured values $T_S = 10$ h and $T_{G_2+M} = 4$ h are used, and it is assumed that $T_C = 20$ h. The ever-present variations in the length of the cycle phases are neglected, again for simplicity of representation; the degree to which the shape of the curve is influenced by such variations will be discussed later.

The theoretical curve (solid line in Fig. 7) begins at $t = 4$ h ($t = 0$ h: time of ^{3}H-TdR injection) with the production of three cell groups with different labels. At this time v is calculated in accordance with Eq. (3):

$$v_{t=4h} = \frac{t}{T_S - \Delta t} \tag{3}$$

(where Δt is the interval between the first injection of ^{3}H-TdR and the second injection of ^{14}C-TdR), i.e., 4 h/10 h - 4 h = 0.67. Equation (3) can be used because the lengths of the cell cycle phases are proportional to the number of cells in these phases (cf. the determination of T_S in Sect. 4.1.4.2). In the first place only the number of purely ^{3}H-labeled cells increases with time t because they first enter mitosis. After a period corresponding to $T_{G_2+M+\Delta t}$ (i.e., after 8 h), the number of these cells had doubled, and with it the value of the ratio v. After this the double-labeled cells also start to undergo mitosis within a time period corresponding to $(T_S - \Delta t)$, or 6 h. Consequently v decreases during this 6-h period to its original value of 0.67. Of course this occurs only if no additional ^{14}C-TdR has been administered. Under this assumption (shown by the dotted line in Fig. 7) the value of v remains constant during the next 10 h (during which time the cells pass through the S phase), and thereafter increases again when the group of purely ^{3}H-labeled cells undergoes the second mitosis after labeling.

As mentioned earlier, an additional injection of ^{14}C-TdR influences the group of purely ^{3}H-labeled cells, and thus the value of v, only if these cells are in the S phase. For a measured T_S of 10 h and measured T_{G_2+M} of 4 h, and an assumed cycle time $T_C = 20$ h, this is the case in the interval between $t = 10$ h and $t = 24$ h. Accordingly, an injection of ^{14}C-TdR during this interval leads to a decrease in the number of purely ^{3}H-labeled cells and hence to an increase in the number of double-labeled cells and a decline in the value of v. In the interval between $t = 14$ h and $t = 20$ h, during which all purely ^{3}H-labeled cells which continue to proliferate aquire a double label, v decreases to 0.24 (instead of 0.67 without a second injection of ^{14}C-TdR).

A comparison between the experimental data (each point signifies the v ratio measured for one rat) and the theoretical curve in Fig. 7 (solid line) shows good agreement. In Fig.7, $T_C = 20$ h was the only unproved assumption made. This agreement is not obtained when theoretical curves are constructed on the assumption that $T_C = 17$ h or $T_C = 23$ h, however. This shows that the investigated glial cells indeed have a T_C of approximately 20 h.

The range of variation of T_C cannot be measured by the method of labeled S phases. At present there is only one experimental method for determining the range of variation of cycle phases; this is a special modification of the method of labeled mitoses (Quastler and Sherman 1959). This method is based on a comparison of the waves of purely ^{3}H- and purely ^{14}C-labeled mitoses after double labeling with ^{3}H-TdR and ^{14}C-TdR (Schultze et al. 1979). However, it is quite possible to obtain a rough esti-mate on the variance of T_C by employing the method of labeled S phases. For example, it can be concluded from Fig. 7 that there are probably few glial cells with $T_C \geq 30$ h; a cycle time of 30 h or more would yield the following theoretical curve (all other assumptions remaining unchanged): Within the first 10 h after ^{3}H-TdR injection the curve would follow a course corresponding to the solid line. Between t = 10 h and t = 20 h, however, it would be identical with the broken line. The further course of the theoretical courve will not be discussed here. This example makes it clear that a large range of variation of T_C, reflected in the presence of many glial cells with $T_C > 30$ h, would lead to values of v which would lie between the solid and broken curves in the interval between t = 12 h and t = 20 h. Before attempting a more precise determination of the variance of T_C from a comparison of the ex-perimental values and theoretical curve, it must be borne in mind that the course of every theoretical curve is naturally dependent on other factors as well, such as variations in T_S and T_{G_2+M}. However, such variations would not have a very great impact on the shape of the curves.

4.2 Methods of Calculating Cell Cycle Parameters

4.2.1 Mean Cycle Time

Besides the methods described previously in Sect. 4.1, the duration of the cell cycle has also been determined for some time solely on the basis of calculations. This computational technique is based on values which are usually easy to measure experimentally, such as the labeling index (LI) and T_S. Of course it must be con-sidered that the labeling index derived from autoradiographs (LI_m), i.e., the portion of labeled cells in relation to all labeled and unlabeled cells of the type investigated, usually refers to proliferating as well as nonproliferating cells. However, as will be shown, the labeling index of the proliferating cells (LI_{GF}) (GF, growth fraction) is necessary for the calculation of T_C. But LI_{GF} and LI_m are identical only if all cells present are proliferating (GF = 1, see Sect. 4.2.3).

Based on the frequency distribution of the cells throughout the cycle (e.g., Lennartz and Maurer 1964) the following relations are obtained between the labeling index of the proliferating cells (LI_{GF}) and the intervals T_C and T_S, depending on the mode of growth (exponential or steady state):

$$LI_{GF} \text{ (steady state)} = T_S/T_C, \tag{4a}$$
i.e., $T_C = T_S/LI_{GF}$ (steady state)

$$LI_{GF} \text{ (exponential)} = 2^{T_{G_2+M}/T_C} (2^{T_S/T_C} - 1) \tag{4b}$$

As mentioned in Sect. 1, estimates of T_C for glial cells in the adult mouse brain have already been made with the aid of Eq. (4a). But it was assumed in such cases that the labeling index derived from autoradiographs, LI_m, was identical with LI_{GF}, i.e., that all glial cells were proliferating — an assumption which is entirely hypothetical for tissues with $LI_m \ll 1$.

When expression (4b) is used to determine T_C, T_{G2+M} must also be known, in addition to LI_{GF} and T_S. However, T_C can be determined with sufficient accuracy even if only rough estimates are available for T_{G2+M}, since T_{G2+M} does not play a significant role in the calculation (cf. Lennartz et al. 1968). Equation (4b) was used by von Waechter and Jaensch (1972) and Schultze et al. (1974a,b) to calculate T_C for neural epithelial cells in the brain of embryonic rats.

4.2.2 Turnover Time

For tissues which are active mitotically but in which the total number of cells remains more or less constant with time, Leblond and Walker (1956) introduced the term "turnover time" (T_T) to describe the time taken for the replacement of a number of cells equal to that in the whole population. T_T thus indicates the mean lifetime of a cell in the population. The turnover time is calculated either from the observed number of mitoses according to Eq. (5a):

$$T_T = T_M/MI \tag{5a}$$

(where T_M is the mitotic time and MI the mitotic index), or from the number of labeled interphase cells observed autoradiographically after injection of ^{3}H- or ^{14}C-TdR according to Eq. (5b):

$$T_T = T_S/LI_m \tag{5b}$$

(where LI_m is the labeling index determined directly from the autoradiographs).

Most calculations of T_T according to Eq. (5b) have been performed for the cells of the subependymal layer as well as for the matrix cells of the cerebellar external granular layer (survey: Lewis 1979). Because the concept of turnover time is meaningful for the various cell types in the brain in only a few cases, no data on T_T were included in the compilation of cell-kinetic data which follows in Sect. 4.3.

4.2.3 Growth Fraction

Mendelsohn (1960, 1962) showed for solid tumors in mice that only a portion of the tumor cells are mitotically active. He called this portion the growth fraction (GF). The nonproliferating cells, collectively termed the nongrowth fraction, are also referred to as cells in the "G_0-phase" (Lajtha 1963; Quastler 1963; see also Epifanova and Terskikh 1969; Burns and Tannock 1970; Smith and Martin 1973; DeMaertelaer and Galand 1975).

In purely formal terms, the term "growth fraction" can also be applied to cells in the brain, e.g., glial cells in the adult mouse brain, by considering the proliferating glial cells scattered throughout the individual brain areas as a unit and relating the number of proliferating glial cells to the total number of glial cells present in a given case.

A direct determination of the growth fraction from autoradiographs is not possible. However, an indirect means is available for determining the growth fraction according to Eq. (6) from the quotient of two known quantities: the measured labeling index (LI_m) and the labeling index of the growth fraction already mentioned in Sect. 4.2.1 (LI_{GF}; see Eqs. 4a and 4b):

$$GF = LI_m / LI_{GF} \qquad\qquad (6)$$

4.3 Compilation of Cell Cycle Parameters

Cell cycle parameters previously published on cell types in the rat and mouse brain in pre- and postnatal ontogeny are compiled in Table 1. Reports which contain only the LI have not been included.

As Table 1 shows, the published data pertain entirely to cells derived from the neural epithelial cells, except for data on endothelial cells in the brain of 14-day-old rats and adult mice. In particular, there is a lack of kinetic studies on cells of meso-dermal origin in the embryonic brain.

To facilitate the comparison of individual data as well as the recognition of trends during ontogeny, the data from Table 1 have been plotted in Fig. 8 for T_S (top) and T_C (bottom) as a function of the age of the animal. No distinction is made between values for rats and mice. As Fig. 8 shows, the individual values for different cell types agree quite well at every age. The data of Shimada (1966) for cells of the subependymal layer, with T_C = 63 h or 65 h, do not conform to the general pattern, but this may be due to methodological inadequacies (determination of T_C from conti-nuous infusion experiments with ^{3}H-TdR; cf. Sect. 4.1.2).

Shortly after formation of the neural tube, T_C equals about 8 h and T_S, about 5 h. These periods become longer with increasing embryonic age. Finally, toward the end of fetal development the cell cycle time is between 18 h and 20 h, and the S phase length is between 7 h and 10 h. These periods then remain largely constant from the end of fetal development on: The mean values for the 51 data points shown between P 1 and adulthood (disregarding the aforementioned values of Shimada (1966) are as follows: T_C = 18.7 h ± 0.5 h and T_S = 9.75 h ± 0.25 h ($\bar{x}$ ± SEM). Similar values of T_C and T_S were also found for Schwann cells in the sciatic nerve of 2-day-old mice (Asbury 1967) as well as for experimentally-induced gliomas in the rat brain (Wilson et al. 1972).

Cell cycle parameters were measured for glial and andothelial cells in four different areas of the brain in 14-day-old rats (Korr 1978a; Korr et al., to be publish-ed b). Various numbers of cells are proliferating in these four areas at any given time. As Table 1 shows, the values found for the labeling index varied by up to a factor of 3. Nevertheless, the same duration was measured in each case for the individual cycle phases. This finding suggests that the same cycle parameters are also valid for the other proliferating glial and endothelial cells present in varying numbers in the individual brain regions (see Mares 1975). Further evidence for the assump-tion that the duration of the cell cycle is independent of the number of proliferating cells is offered by studies on Schwann cells in the adult mouse sciatic nerve following neurotomy (Bradley and Asbury 1970).

The growth fraction in the ventricular layer of the neural tube is nearly equal

Table 1. Compilation of cell cycle parameters for various cell types in the rat and mouse CNS during the course of pre- and postnatal ontogeny

I	II	III[a]	IV				V		VI
Cell type	Animal	Age	T_C (h)	T_S (h)	T_{G_2+M} (h)	T_{G_1} (h)	LI (%)	GF	References
Neural tube	mouse	E10	7	5.1	1.8	0.1			Hoshino et al. (1973)
	mouse	E10	8.5	4.0	2.2	2.3	50.6	1.0[b]	Kauffman (1966)
	mouse	E10	8.5	4.6	~2	~1.9	53	1.0[b]	Kauffman (1968)
	mouse	E11	10.5	5.4	2.4	2.7	49	1.0[b]	Kauffman (1968)
	mouse	E11	~11	5.5	~2	~3.5	50	1.0[b]	Atlas and Bond (1965)
	rat	E12	13.1	6.9		3.7	50.5		Waechter and Jaensch (1972)
	rat	E13	14.2	7.4		4.3	49.5		Waechter and Jaensch (1972)
	mouse	E13	15.5	6.9	1.8	6.8			Hoshino et al. (1973)
	rat	E14	12.9	6.9		3.5	50.9		Waechter and Jaensch (1972)
	rat	E14	~12						Ellenberger et al. (1969)
	rat	E15	16.3	6.8	2.0	7.5	39.8	0.97	Gracheva (1964)
	rat	E15	18.4	9.4		6.5	46.8		Waechter and Jaensch (1972)
	mouse	E15	~11	7.5	~2				Langman and Welch (1967)
	rat	E16	17.0	7.7		6.8	40.9		Waechter and Jaensch (1972)
	mouse	E16	18.5	6.5	3.2	8.8	36±3.6	1.0[b]	Shimada et al. (1977)
	rat	E10-16	10	6			60	1.0	Schultze et al. (1974)
	rat	E17	16.8	6.8		7.5	35.3		Waechter and Jaensch (1972)
	mouse	E17	26	10.4	1.8	13.8			Hoshino et al. (1973)
	rat	E18	20	5.5	2.5	12	19.8	0.65	Gracheva (1964)
	rat	E18	20.8	7.1		11.2	28.9		Waechter and Jaensch (1972)
	rat	E20	18	5.5	2.5	10	15.1	0.50	Gracheva (1964)
	rat	P1	26	11.7	2.7	11	7.7	0.18	Gracheva (1964)
Retina	mouse	E10	10	6.5				1.0	Konyukhov and Sazhina (1971)
	mouse	E10/11	~9	~6			75	1.0[b]	Sinitsina (1971)
	mouse	E14	~13.5	~8.5			47.9	0.76[b]	Sinitsina (1971)
	mouse	E15	20	9.8					Konyukhov and Sazhina (1976)
	rat	E15	19.5	10.5	~2	~7	30		Zavarzin and Stroyeva (1964)
	mouse	E16		~10			40.4		Sinitsina (1971)
	rat	P2	28	12.5	2.5	13	15	0.33[b]	Denham (1967)
Cerebellar external granular layer	rat	E15	18	7.5	~1.5	~9	27		Nevmivaka (1964)
	rat	E18	21	5	~2.5	~14.5	39		Nevmivaka (1964)
	rat	P1	26.5	9.5	1.5	15.5	48		Nevmivaka (1964)
	rat	P1/2	19.0	9.2	~1.5	8.3	25.7	0.50	Lewis et al. (1975)
	mouse	P1	29	7.3	2.7	19	30	1.0	Fujita et al. (1966)
	mouse	P2	14.5	8.0				0.83	Mares and Lodin (1970)
	mouse	P2	16.2	7.4				0.79	Mares and Lodin (1970)
	mouse	P2	16	7.3			45	0.83	Mares et al. (1970)
	mouse	P3	21	7	2.5	11.5	41	1.0	Fujita et al. (1966)
	rat	P6	16.3	9.6	2.5	4.2	24.7	0.42	Lewis et al. (1976)
	rat	P6/7	16.3	9.6	~2.5	4.2	27.2	0.47	Lewis et al. (1975)
	mouse	P7	17.6	8.8				0.99	Mares and Lodin (1970)
	mouse	P7	19.3	8.6				0.98	Mares and Lodin (1970)
	mouse	P7	18	8.5			50	0.99	Mares et al. (1970)
	mouse	P7	15	5.6	2.4	7	44	1.0	Fujita et al. (1966)
	mouse	P10	20	9			43	0.80	Mares et al. (1970)
	mouse	P10	21	8.1	2.6	10.3	43	1.0	Fujita et al. 1966)
	rat	P10	19.8	9.0	2.5	8.3	22	0.48	Lauder (1977)
	mouse	P10/11	19	8	2.5	8.5	43	1.0	Fujita (1967)
	rat	P11	17.1	11.1	~2.3	3.7	23.5	0.36[b]	Lewis et al. (1977 c)
	rat	P12	16.8	10.6	2.4	3.8	21.9	0.34	Lewis et al. (1976)
	rat	P12/13	16.8	10.6	~2.4	3.8	21.9	0.34	Lewis et al. (1975)
	rat	P21	17.5	11.2	3.3	3.0	17.2	0.31	Lewis et al. (1976)
	rat	P21/22	17.3	9.7	~2.8	4.8	17.2	0.31	Lewis et al. (1975)
Hippocampal dentate gyrus	rat	P1	15.1	10.1	~3.9	1.1	4.5	0.07[b]	Lewis (1978)
	rat	P6	17.7	11.7	~3.6	2.4	3.7	0.06[b]	Lewis (1978)
	rat	P12	15.3	11.2	~2.9	1.2			Lewis (1978)
Subependymal layer (forebrain lateral ventricles)	rat	P1	15.5	5.7			21.5	0.58[b]	Knowles (1976)
	rat	P1	26	11.7	2.8	11.5	7.7	0.17[b]	Gracheva (1969)
	rat	P1	18.3	10.0			12.6	0.23[b]	Lewis and Lai (1974)
	rat	P1/2	18.3	10.0	~5.2	3.1	12.6	0.24	Lewis et al. (1975)
	mouse	P1	63	7.0		54			Shimada (1966)

I	II	IIIa	IV				V		VI
Cell type	Animal	Age	T_C (h)	T_S (h)	T_{G_2+M} (h)	T_{G_1} (h)	LI (%)	GF	References
	mouse	P3	65	10.0		53			Shimada (1966)
	rat	P6	17.2	10.8			9.7	0.15[b]	Lewis and Lai (1974)
	rat	P6	17.2	10.8	3.9	2.5	9.7	0.16	Lewis et al. (1976)
	rat	P6/7	17.2	10.8	~3.9	2.5	9.7	0.16	Lewis et al. (1975)
	rat	P11	14.7	11.8	~2.5	0.4	10.2	0.13[b]	Lewis et al. (1977 c)
	rat	P12	15.3	10.9			9.3	0.13[b]	Lewis and Lai (1974)
	rat	P12	15.3	10.9	2.5	1.9	9.3	0.14	Lewis et al. (1976)
	rat	P12/13	15.3	10.9	~2.5	1.9	9.3	0.14	Lewis et al. (1975)
	rat	P20	18	5.5	2.5	10	15.1	0.49[b]	Gracheva (1969)
	rat	P21	20.1	12.4			11.8	0.19[b]	Lewis and Lai (1974)
	rat	P21	19.7	12.8	3.9	3.0	11.8	0.19	Lewis et al. (1976)
	rat	P21/22	20.1	12.4	~2.5	5.2	11.8	0.19	Lewis et al. (1975)
	rat	adult	18	8.5	~2	~7.5	16.6	0.35[b]	Lewis (1968 a)
	rat	adult	21	9.8	3.8	7.4	7.5	0.16[b]	Gracheva (1969)
	rat	adult	20.4	12.3	~2.7	~4.3	12.4	0.21[b]	Lewis et al. (1977 b)
	mouse	adult	< 25				19.3	0.40	Korr (1978 a)
Glial cells Cerebellum	rat	P6	19.5	8.6	~3.4	7.5	9.7	0.22[b]	Lewis et al. (1977 a)
	rat	P6/7		6			12	0.25	Moskovkin et al. (1978)
	rat	P7		10.5	~3		35		Moskovkin (1976)
	rat	P7		10	~3		55		Moskovkin (1976)
	rat	P12	14.3	9.2	~4	1.1	0.73	0.011[b]	Lewis et al. (1977 a)
Cortex	rat	P14	20	10.1±0.6	~4	~6	1.8±0.2	0.04	Korr (1978 a)
Corp. callosum	rat	P14	20	10.1±0.7	~4	~6	5.4±0.8	0.11	Korr (1978 a)
N. caudatus	rat	P14	20	10.0±0.6			3.0±0.3	0.06	Korr (1978 a)
Comm. anterior	rat	P14	20	10.2±0.6			5.2±0.4	0.11	Korr (1978 a)
Forebrain	mouse	adult	20	9.4±0.5	~5	~5	0.2	0.004	Korr et al., (1973, 1975)
Astrocytes			20	~10	~5	~5			
Oligodendr.			20	~10	~5	~5			
Forebrain	rat	adult			4-5				Hommes and Leblond (1967)
Endothelial cells Cortex	rat	P14	20	9.9±1.0			2.7±0.8	0.06	Korr (1978 a)
Corp. callosum	rat	P14	20	9.4±1.3			3.2±0.4	0.08	Korr (1978 a)
N. caudatus	rat	P14	20	9.4±1.1			2.4±0.3	0.05	Korr (1978 a)
Comm. anterior	rat	P14	20	8.9±2.2			5.3±2.0	0.12	Korr (1978 a)
Forebrain	mouse	adult	20	10.6±1.7			0.14	0.003	Korr (1978 a)

[a] E, embryonic day after fertilization; P, postnatal day.
[b] The growth fraction (GF) was computed from the values given in column IV for T_S, T_C and the labeling index (LI, column V) in accordance with Eqs. (4a) and (6).

to 1.0 up to about fetal day 16, i.e., all cells are involved in proliferation. These conditions change with time. This is shown very clearly by studies on comparable regions of the embryonic brain between E 15 and P 1 (Gracheva 1964): The growth fraction declines sharply toward the end of fetal development. A decline in the growth fraction with age similar in degree to that occurring in the ventricular layer of the neural tube is also found in the cerebellar external granular layer.

The growth fraction of the cells of the subependymal layer exhibit a marked range of variation in comparison with the other cell types of the brain. These variations are caused mainly by different labeling indices. It is possible that the studies were done on different regions of the subependymal layer. Gracheva (1969), for example, showed that the mitotic index for subependymal cells of the olfactory bulb as well as the lateral ventricle of the forebrain in the adult rat varies by a factor of about 4 to 15. No dependence of the growth fraction of the subependymal cells on the age of the animal can be discerned from the data in Table 1. As our own

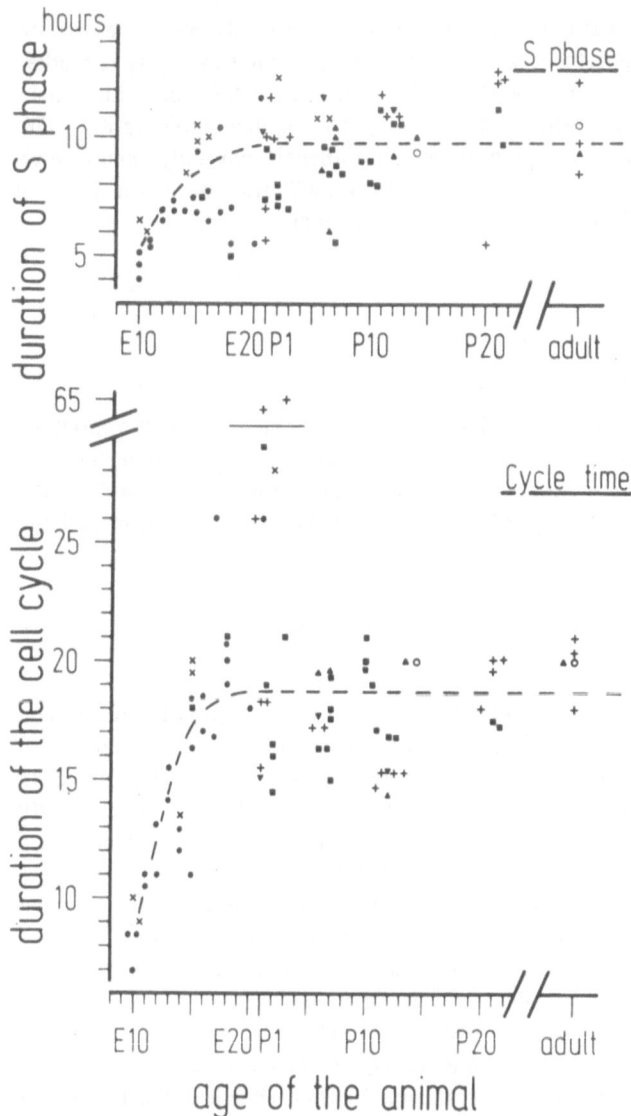

Fig. 8. Length of S-phase (T$_S$; *top*) and cell cycle time (T$_C$; *bottom*) of various cell types in the brain of the rat and mouse as a function of age. Symbols: ●, ventricular cells of the neural tube; x, matrix cells of the retina; ■, cells in the external granular layer of the cerebellum; +, cells in the subependymal layer of the lateral ventricle; ▲, astrocytes and oligodendrocytes; ○, endothelial cells

studies in 6- to 18-month-old mice have shown (Korr, to be published), the growth fraction of the subependymal cells probably declines markedly in old animals as compared with the values shown in Table 1 for adult rats and mice.

Although the growth fraction of the glial cells in the prenatal brain has not previously been measured, the values are probably similar to those of their precursors (i.e., the neural epithelial cells). This means that the growth fraction of the neuroglia is approximately 1.0 during early embryonic development and decreases with age without reaching zero. The same is probably also true for the endothelial cells. Accordingly, the course of pre- and postnatal ontogeny is marked by a steady decline in the number of those glial and endothelial cells which make the entry into the S phase which is necessary for further proliferation. This also means that in the case of the glial and endothelial cells only quantitative, but not qualitative, differences exist between the growth fraction in the brain of a very young animal and that in the adult or very old animal.

5 Mode of Proliferation

Having determined the cell cycle parameters, we shall next examine the mode of proliferation of the individual cell types. Our studies have shown that the proliferation of various cell types occurs on the one hand with cell loss, and on the other with a permanent interchange of cells between the growth fraction and the non-growth fraction.

5.1 Proliferation with Cell Loss

In the studies that follow, pyknotic cells play a decisive role. Pyknotic nuclei, i.e., the nuclei of degenerating cells which are subsequently phagocytized, have been known in the CNS for a long time. The presence of pyknotic cells is particularly marked in embryonic brains (for survey see Glücksmann 1951; Sidman 1970; Prestige 1974; Silver 1976), as well as after X-irradiation in the brains of young animals (e.g., Altman and Nicholson 1971). In the latter case the pyknoses were found primarily in areas of very active proliferation such as the cerebellar external granular layer. Moreover, pyknotic glial cells and pyknotic cells of the subependymal layer have also been found in small numbers in the brain of untreated animals of various species (Pannese and Ferrannini 1967). These pyknoses were originally regarded as fixation artifacts resulting from prior immersion fixation. However, this assumption became untenable when it was learned that pyknoses are also observed after careful perfusion fixation (Pannese and Ferrannini 1967; Sturrock 1974d). Thus pyknoses, like mitoses, are included among the cell forms which normally occur in the animal brain.

The observation of both degenerating cells and mitoses in the glial cells of the adult rat and mouse brain was interpreted as evidence of a turnover of these cells. Accordingly, pyknotic glial cells were regarded as aged cells which die and are replaced by newly formed glial cells. It will be shown below, however, that pyknoses arise not only from aged cells, but also from proliferating cells which have just undergone mitosis. In this way connections which previously were only presumed to exist between cell proliferation and the occurrence of pyknoses (Cammermeyer 1970; Lewis 1975) can be experimentally demonstrated.

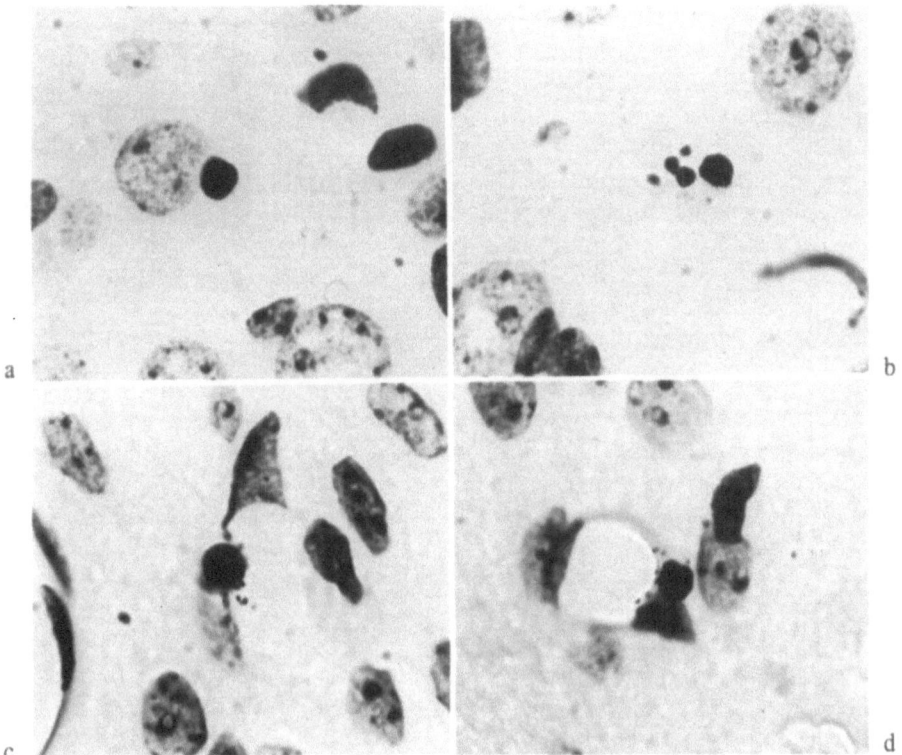

Fig. 9 a–d. Pyknotic nuclei in the cortex of the untreated adult mouse. *a* Pyknotic glial cell in perineuronal position; *b* pyknotic pair of glial cells; one of the two pyknotic nuclei has already disintegrated into three smaller units; *c and d* pyknotic endothelial cells. X 1300

5.1.1 Pyknotic Neuroglial Cells

In autoradiographic studies with ^{3}H- or ^{14}C-TdR, unlabeled and in many cases labeled pyknotic glial cells were found in practically all autoradiographs from the forebrain of adult, untreated mice (Korr et al. 1973, 1975; Korr 1978a). These pyknotic nuclei, like those in embryonic brains or irradiated brain tissues, are very densely stained by the Feulgen method and usually appear circular in the section. In contrast to mitotic nuclei (e.g., telophase nuclei), their karyoplasm shows no structure. The chromatin of the nucleus appears as a homogenous sphere. Figure 9a shows a typical example of a pyknotic glial cell.

Besides isolated pyknotic glial cells as shown in Fig. 9a, labeled and unlabeled pairs of pyknotic glial cells were also observed. These pyknoses apparently arose from glial cell pairs, i.e., glial cells which have undergone mitosis and are still in close spatial proximity (cf. Hommes and Leblond 1967, on the observation of glial cell pairs). The two pyknotic nuclei observed in such pairs were either both labeled or both unlabeled. No case was observed in which only one of the two pyknotic nuclei was labeled.

As various studies have shown, pyknotic cells in the brain become karyorrhectic and are finally phagocytized (Altman and Nicholson 1971; Sturrock 1974d; Lewis, 1975). Even though the phagocytosis of cells with karyorrhectic nuclei could not be

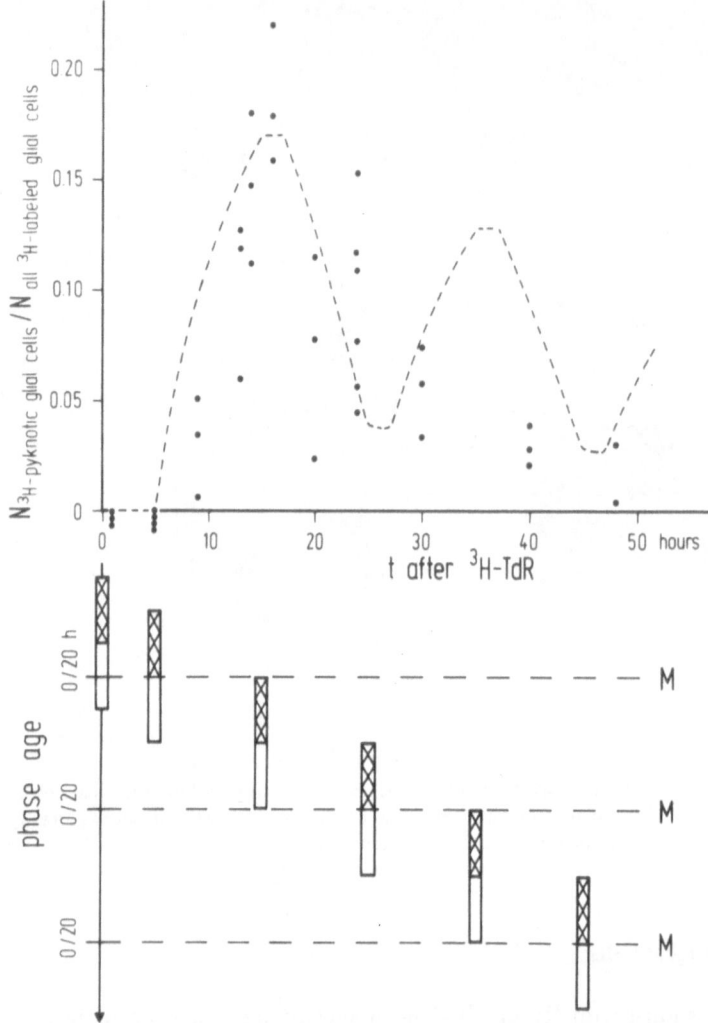

Fig. 10. Ratio of ^{3}H-labeled pyknotic glial cells to all ^{3}H-labeled glial cells in the brain of the adult mouse as a function of time after ^{3}H-TdR injection. *Top,* experimental values for individual animals *(dots)* and theoretical curve *(broken line)*. *Bottom,* schematic correlation of phase diagrams of labeled *(shaded rectangle)* and unlabeled *(unshaded rectangle)* proliferating glial cells at time *t* after ^{3}H-TdR injection

demonstrated by Feulgen staining in the studies discussed here on the adult mouse brain, the following observations provide evidence that the pyknotic glial cell nuclei actually represent degenerate cells, and thus that pyknoses are a manifestation of cell loss: In many cases clusters of several small pyknotic bodies were found, each about 1-2 μm in diameter. These clusters had about the same area as a pyknotic nucleus shortly after its formation and are suggestive of a decomposing nucleus. Fig. 9 b shows a typical example of this. Here one of the nuclei in a pyknotic glial cell pair has already disintegrated into three smaller units.

In the following we will analyze more closely the numbers of ^{3}H-labeled and (in later sections) unlabeled pyknotic glial cells found in various experiments. Figure 10 (top) shows the numbers of labeled pyknotic glial cells determined. The experimental data are plotted as relative values, i.e., as the ratio of the ^{3}H-labeled pyknotic glial cells to all ^{3}H-labeled glial cells as a function of time after ^{3}H-TdR injection. As Fig. 10 shows, no labeled pyknotic glial cells were observed 1 h to 5 h after ^{3}H-TdR injection. The first labeled pyknotic glial cells appeared after 5 h, and their relative number increased until about t = 14-16 h. Later a decline was observed. Besides these labeled pyknotic glial cells, unlabeled pyknoses were always found.

This varying distribution of labeled and unlabeled pyknotic glial cells, together with the previous argument concerning the detection of pyknoses after both immersion and perfusion fixation, offers convincing evidence that pyknoses generally cannot be interpreted as fixation artifacts.

5.1.1.1 Formation of Pyknotic Glial Cells from Proliferating Glial Cells

a) Formation of ^{3}H-labeled pyknotic glial cells from ^{3}H-labeled glial cells. The observation of *labeled* pyknotic glial cells suggests that a connection exists between the formation of pyknotic cells and glial cell proliferation. This is also evidenced by the finding of labeled as well as unlabeled *pyknotic glial cell pairs.* Their occurrence indicates that pyknosis develops after the cell has undergone mitosis, probably in the early G_1 phase. Evidence for this assumption is provided by the mean grain count of the labeled pyknotic nuclei in the period from t = 9 h to t = 24 h, which corresponds to that of glial cells which have already undergone mitosis. Thus, the labeled pyknotic glial cells were formed at a time when the labeled glial cells had already reduced their grain count to half its value, i.e., after the completion of a mitosis.

The time of existence within the brain of the labeled pyknotic nuclei can be estimated from the comparison of the mean grain count of the labeled pyknotic glial cells at different times after ^{3}H-TdR injection: In the time interval between t = 30 h and t = 40 h apparently no labeled pyknotic nuclei can be observed that were present in the interval from t = 9 h to t = 24 h, because the associated mean grain counts follow an approximate 0.5:1 ratio. Compared with t = 15 h, i.e., the time at which theoretically all labeled glial cells have undergone mitosis and thus the last labeled pyknotic glial cells after this mitosis have formed (see below), we thus obtain a minimum time of existence within the brain of more than 9 h (t = 15 h to t = 24 h) and a maximum time of existence within the brain of less that 15 h (t = 15 h to t = 30 h), or a mean value of about 12 h.

The connection previously described between pyknotic glial cells and glial cell proliferation can be represented by a *theoretical curve* (broken line in Fig. 10). For this curve the cell cycle parameters of the glial cells of the animal under study must be known; i.e., the formation of pyknotic glial cells is correlated with the phase of the proliferating glial cells (Fig. 10, bottom): If labeled pyknoses form at the start of the G_1-phase from proliferating cells which have undergone mitosis, the first labeled pyknoses should be encountered after a period which is somewhat longer than the duration of the phases (G_2 + M), i.e., somewhat more than 5 h after ^{3}H-TdR injection. If labeled pyknotic glial cells are observed for a period longer than a few hours after their formation, the percentage of labeled pyknoses should increase with time until the point at which all labeled cells have undergone mitosis. Assuming constant cycle times and a time of existence of at least 10 h for the pyknotic nuclei, the labeled pyknoses would reach a peak after a period corresponding to the duration of the phases ($S + G_2 + M$), or 15 h. If the formation of pyknotic nuclei is not a singular event, labeled pyknoses should again arise after furter mitoses are completed.

An assumption must also be made regarding the degree to which pyknotic nuclei are formed. This was done on the basis of a measured maximum ratio of labeled pyknotic glial cells to all ^{3}H-labeled glial cells of 0.17 (mean value of ratios in the interval from t = 14 h to t = 16 h). From this maximum value, the length of the S phase (10 h) and the time of existence within the brain of pyknotic nuclei (12 h), it was empirically assumed that for each glial cell labeled at the start of the experiment, 0.034 labeled daughter cells per h become pyknotic. Under this special assumption on the degree of the cell loss and its occurrence within the cell cycle, the dashed curve in Fig. 10 (top) was constructed.

As Fig. 10 shows, the experimental data are in rather good agreement with this theoretical curve. The same is true for the grain counts of the labeled pyknotic nuclei: Thus, it can be deduced from the grain count of the labeled pyknoses found 30 and 40 h after the start of the experiment that these pyknotic nuclei were formed after the labeled glial cells had undergone their second mitosis (from the time of labeling), because they exhibited a mean grain count per nucleus which was half that of those labeled pyknotic glial cells which were found in the interval from t = 9 h to t = 24 h. From the same considerations it can be assumed that the labeled pyknotic nuclei at t = 48 h originated from neuroglial cells which had already completed their third mitosis after labeling.

On the one hand, the agreement between experimental values and the theoretical curve confirms our assumption on the formation of pyknotic glial cells. On the other hand, it provides an estimate of the degree of cell loss associated with proliferation: labeled pyknotic nuclei are produced at a rate of 0.034 labeled daughter cells per hour (the "pyknotic rate"). It must be borne in mind that the measured maximum value of 0.17 obtained from the product of the pyknotic rate and the time of existence within the brain of pyknoses can also be derived from other pyknotic rates and times of existence; e.g., from a pyknotic rate which is twice as high (0.068 instead of 0.034 and a time of existence of the pyknoses which is half as long (6 h instead of 12 h). The theoretical curve predicated upon these assumptions would then no longer coincide with other experimental values. But even though the pair of values used here (i.e., pyknotic rate of 0.034 labeled daughter cells per hour, time of existence within the brain of 12 h) cannot be considered the only solution, a satisfactory estimate is nonetheless obtained: For each glial cell labeled in the S phase, about 0.34 labeled pyknotic glial cells are formed after mitosis, i.e., about 17 % of the labeled daughter cells become pyknotic. This percentage can be considered as cell loss.

b) Formation of unlabeled pyknotic glial cells from proliferating glial cells. If labeled pyknotic glial cells are formed from proliferating, labeled glial cells after mitosis, a similar situation should exist with respect to *unlabeled* pyknoses and unlabeled proliferating glial cells. Our studies in this area will be based on the ratios of the number of unlabeled pyknotic cells to the number of all ^{3}H-labeled glial cells as determined by autoradiographic means. Although, in contrast to labeled pyknotic cells, no direct relation exists in this case between the pyknoses and their cells of origin, the experimental data can be compared with corresponding theoretical values which are based on the assumption that unlabeled pyknoses are formed during the course of cell proliferation. One such theoretical curve is shown as a broken line in Fig. 11 (for details

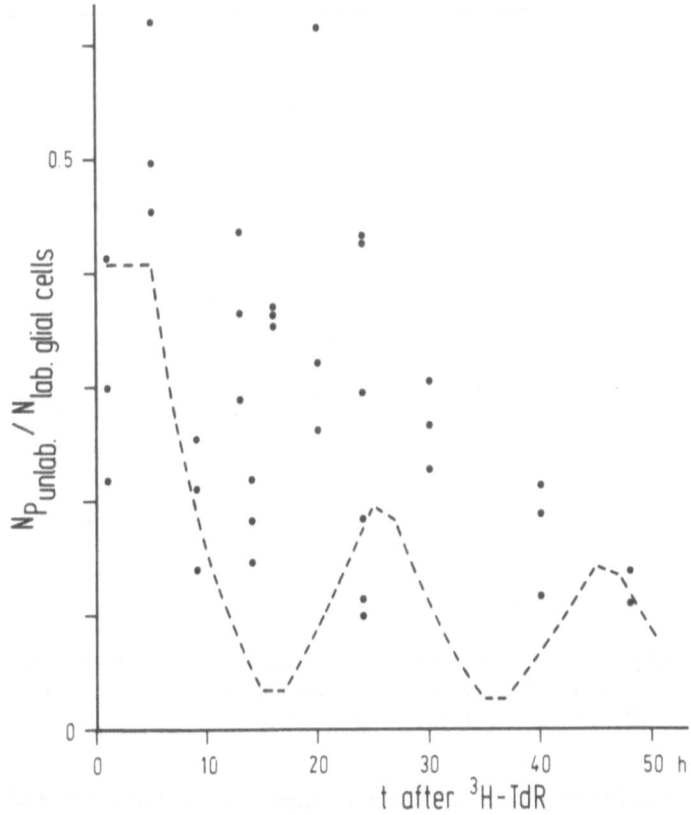

Fig. 11. Ratio of unlabeled pyknotic glial cells to all ^{3}H-labeled glial cells in the brain of the adult mouse as a function of time after ^{3}H-TdR injection. *Dots*, experimental values for individual animals; *broken line*, theoretical curve (see text)

on its construction see Korr 1978a). Figure 11 also shows the experimentally determined ratios of unlabeled pyknotic glial cells to all ^{3}H-labeled glial cells. Each point indicates this ratio for one mouse. As Fig. 11 shows, most of the points lie above the theoretical curve. This suggests that more unlabeled pyknoses are present than would be expected from extimates on the degree of pyknotic glial cell production associated with cell proliferation.

c) Further considerations on the formation of pyknotic glial cells. In another experiment to answer the question of the origin of the pyknotic glial cells, we investigated the ratio of the number of labeled plus unlabeled pyknotic glial cells to the number of all ^{3}H-labeled glial cells as a function of time t after ^{3}H-TdR injection. If unlabeled pyknoses are formed in the same way as labeled pyknoses, this process should follow an alternating pattern after ^{3}H-TdR injection, so that overall the total of all pyknotic glial cells formed should remain constant with time.

Like Figs. 10 and 11, Fig. 12 shows a theoretical curve (broken line) along with experimental data (dots). This curve was based on the assumption that labeled and unlabeled pyknotic glial cells are formed from labeled and unlabeled proliferating glial cells after they have undergone mitosis (for details on construction of the curve see Korr 1978a). As Fig. 12 shows, most of the experimental values lie above the theo-

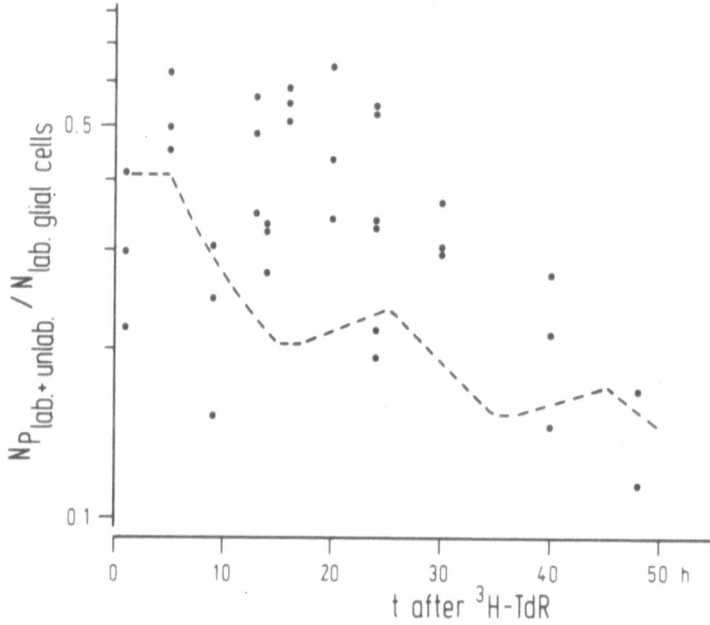

Fig. 12. Ratio of ^{3}H-labeled plus unlabeled pyknotic glial cells to all ^{3}H-labeled glial cells in the brain of the adult mouse as a function of time after ^{3}H-TdR injection. *Dots*, experimental values for individual animals; *broken line*, theoretical curve (see text)

retical curve; however, a satisfactory agreement between experiment and theory would be obtained if the theoretical curve were situated higher above the time axis. Two conclusions can be drawn from this: first, the experimental values are not inconsistent with the assumption that the total number of pyknoses formed from proliferating glial cells remains constant with time; second, unlabeled pyknotic glial cells are apparently formed by another mechanism which is also time-constant: perhaps from aged, non-proliferating glial cells (cf. Sect. 5.1). Although more precise studies are lacking, this offers one means of interpreting the present findings. Accordingly, the experimental values in Figs. 11 and 12 appear to reflect two phenomena: first, the formation of unlabeled pyknotic glial cells from proliferating glial cells at the transition between mitosis and the G_1 phase, and second the (probably continuous) formation of unlabeled pyknoses from aged glial cells. Available numerical data are still too fragmentary for quantitative conclusions to be drawn on the degree of the cell loss from aged glial cells. Nevertheless, it is probable that unlabeled pyknoses are produced from unlabeled proliferating glial cells to the same degree estimated for labeled pyknotic glial cells. On the whole, then, the cell loss can be considered in relation to all proliferating cells; i.e., it appears to be true that about 17 % of all daughter cells become pyknotic following a mitosis and that this percentage can be considered as cell loss.

d) Effects of radiation. The appearance of pyknotic cells following X-irradiation (Altman and Nicholson 1971), or the observation of labeled pyknotic cells in the subependymal layer following the intraventricular injection of large doses of ^{3}H-TdR (Lewis 1968b), raises the question of whether the observed labeled pyknoses may be the result of radiation damage from the uptake of ^{3}H-TdR. However, the connection between labeled as well as unlabeled pyknotic glial cells and proliferating glial cells shows that labeled pyknotic glial cells cannot generally by attributed to radiation damage. Of course it is still possible that some labeled glial cells became pyknotic entirely because of the labeled TdR incorporated into their nucleus.

5.1.1.2 Pyknotic Glial Cells in the Brain of the Young Rat

As in the adult mouse, labeled and unlabeled pyknotic glial cells were also found in the forebrain of the 14-day-old rat. Again, the measured ratios were compared with theoretical curves in the same way as for the adult mouse. Remarkably, all experimental values (the ratios of labeled, unlabeled and, total pyknotic cells related to all ^{3}H-labeled glial cells) were in good agreement with theoretical values, if a cell loss of approximately 3% of the daughter cells was assumed. Moreover, pyknoses apparently do not arise from aged glial cells in these young animals (for details see Korr 1978a; Korr et al., to be published c).

5.1.1.3 Pyknotic Glial Cells in the Brain of Prenatal Rodents

No experimental data are available on the ratio of labeled or unlabeled pyknotic glial cells to ^{3}H-labeled glial cells for the prenatal phase of life. However, the following observations are noted:

Sturrock (1974b,d) described both mitoses and pyknoses of glial cells in the anterior limb of the anterior commissure in mice from fetal day 16 on. At this time the first glial cells in the embryonic brain become visible in this area (Sturrock 1974a). In the human fetus, pyknotic cells are found in the optic nerve, along with astrocytes and astrocyte mitoses, at 8 weeks after fertilization (Sturrock 1975). These observations provide evidence that even prenatally, neuroglial cells can undergo proliferation with cell loss.

In summary, then, it can be stated that the neuroglial cells proliferate with a constant cell loss in the normal, untreated animal during both pre- and postnatal life. It is not known whether differences exist between astrocytes and oligodendrocytes in this regard. In the 14-day-old rat about 3% of the daughter cells following a mitosis become pyknotic and die; in the adult mouse glial cell loss is about 17%. This suggests that the cell loss associated with proliferation increases with the age of the animal. In adult animals, moreover, cell loss appears to occur from aged, nonproliferating glial cells.

5.1.2 Pyknotic Cells of the Subependymal Layer

Various authors have described pyknotic cells in the subependymal layer of untreated young and adult rats and mice (Smart 1961; Lewis 1968b; Privat and Leblond 1972; Lewis and Lai 1974; Lewis 1975; Lewis et al. 1975, 1976, 1977a,b,c; Knowles 1976; Korr 1978a). Besides unlabeled pyknoses (pyknotic indices are reported in the papers of Lewis and Knowles), observations of labeled pyknoses following ^{3}H-TdR injection are also described. Interestingly, labeled pyknotic nuclei did not appear immediately after ^{3}H-TdR injection, but only after a delay of 3-4 h (Lewis 1975; Knowles 1976; Korr 1978a). This suggests that these labeled pyknoses are formed from proliferating cells after mitosis in a manner similar to that demonstrated in Sect. 5.1.1.1 for the glial cells in the adult mouse brain. Thus, the cells in the subependymal layer apparently proliferate with cell loss.

A rough estimate of the degree of cell loss associated with proliferation in the subependymal layer of the lateral ventricle of adult mice is obtained as follows (cf. Korr 1978a): 14 h after injection of ^{3}H-TdR, a value of about 0.09 was measured for the ratio of ^{3}H-labeled pyknoses to all ^{3}H-labeled subependymal cells. In the case of the neuroglia, the rate of formation of pyknoses relative to the proliferating cells was satisfactorily estimated from a ratio measured at a similar point in time, i.e., when all cells labeled in the S phase had undergone the next mitosis. This estimate was based upon the known length of the S phase and an assumed time of existence of the pyknotic nuclei in the brain $\geqslant T_S$. If the time of existence of the pyknotic nuclei in the subependymal layer is between 9 and 15 h as with the pyknotic glial cells, it can be assumed (analogous to the neuroglia) that approximately 9% of the daughter cells produced by a mitosis in the subependymal layer of the adult mouse brain will become pyknotic, and can be considered as cell loss.

5.1.3 Pyknotic Precursor Cells of Neurons

Various authors have reported numerous pyknotic cells in the embryonic brain even under normal conditions (survey: Glücksmann 1951; Sidman 1970; Prestige 1974; Silver 1976). These pyknoses are reported to occur mainly during the differentiation phase of the neuroblasts. The question of what caused the cell death remains unanswered (cf. Prestige 1974). However, one must consider the possibility that at least some of the pyknoses were produced in the same way shown for neuroglial cells in the brain of the young rat and adult mouse, i.e., in connection with the proliferation of neuroepithelial cells. This is also suggested by the report of Sidman (1970), who found labeled pyknotic cells in the neuroepithelium of the retina of 3-day-old mice 20 h after injection of ^{3}H-TdR.

Lewis (1975) also found first unlabeled and then labeled pyknotic cells in the external granular layer of the cerebellum in young rats in the interval from 1 h to about 4 h after ^{3}H-TdR injection. As studies by Swarz and del Cerro (1977) have shown, cells produced in the external granular layer later develop into neurons. We thus have an example where the immediate precursor cells of neurons have a mode of proliferation which is similar to that of neuroglial cells and cells of the subependymal layer, namely proliferation with cell loss.

5.1.4 Pyknotic Endothelial Cells

Pyknotic endothelial cells in the brain of young and adult rodents have only been rarely described (Korr et al. 1975; Korr 1978a). This may have to do with the fact that a pyknotic cell in a paraffin section can be identified as an endothelial cell only if the associated capillary is also visible in cross section. Figure 9c,d shows two examples of labeled pyknotic endothelial cells from the cortex of the adult mouse.

In the case of endothelial cells, it is more difficult to demonstrate a connection between proliferating cells and the formation of pyknoses than in the case of the glial cells. Nevertheless, the observation of labeled and unlabeled pyknotic endothelial cells suggests that proliferation of endothelial cells is also connected with cell loss.

5.2. Proliferation with Exchange of Cells between the Growth Fraction and Nongrowth Fraction

In the preceding chapters the proliferation of certain cell types of the CNS was characterized in some detail, first by the indication of cell cycle parameters and second by showing that proliferation is accompanied by cell loss. In the present chapter we shall examine the behavior of the cells of the growth fraction and nongrowth fraction as revealed by autoradiographic studies of neuroglial cells in the forebrain of the adult mouse. We will also consider the extent to which the relationships found for the glial cells can be applied to other cell types of the CNS in the course of pre- and postnatal ontogeny.

5.2.1 Neuroglia

5.2.1.1 Studies of the Neuroglia in the Adult Mouse Brain

a) Passage of glial cells from the growth fraction to the nongrowth fraction. As indicated in Table 1, the labeling index of the neuroglial cells in the forebrain of adult mice is 0.2%; thus, only a few labeled glial cells can be observed after injection of ^{3}H-TdR. Now all these labeled glial cells cannot continue to proliferate, or else a tremendous number of new glial cells would be produced within a short time. But available histometric data on changes in glial cell density with age (Brizzee et al. 1964; Ling and Leblond 1973; Bayer 1977) indicate that this is not the case. Because the growth fraction of the glial cells in the adult animal brain remains more or less constant over periods of several weeks, it can be concluded that of the two daughter cells produced by a mitosis, only one continues to proliferate, while the other daughter cell stops proliferating, i.e., passes into the nongrowth fraction. Thus, all considerations point to the fact that the glial cells undergo a steady state growth in the brain of adult mice.

These considerations can also be confirmed experimentally: Following a single injection of ^{3}H-TdR or ^{14}C-TdR, a great many labeled glial cells were found 6,8 or 14 days later (Korr 1978a). These labeled glial cells must have previously left the growth fraction, or else their grain count would have been too highly diluted in relation to the grain count shortly after injection of labeled TdR.

An attempt was also made to demonstrate the passage of glial cells into the nongrowth fraction in quantitative terms, assuming a steady state mode of growth. For this purpose the experimental values – i) number of ^{3}H-labeled glial cells per brain section, ii) mean grain count per nucleus of the ^{3}H-labeled glial cells, and iii) the frequency distribution of the grain count per nucleus – were analyzed more closely. Without going into details (see Korr 1978a), we shall present the essential features of these studies:

i) The numbers of ^{3}H-labeled glial cells per brain section (forebrain between the sectional plane of corpus callosum/anterior commissure and corpus callosum/commissure of ventral fornix; cf. Fig. 1 in Korr et al. 1973) are plotted on a semilog scale in Fig. 13 against time after ^{3}H-TdR injection. Each point in Fig. 13 corresponds to the value measured for one mouse. As is seen, the individual values fluctuate greatly at the start of the experiment; values for different experimental animals vary by as much as a factor of 2. Nevertheless, an increase from about 42 labeled glial cells per brain section to about 238 labeled cells per section is clearly evident. The former number corresponds to the mean value for the period of 1–5 h after ^{3}H-TdR injection, i.e., for that

44

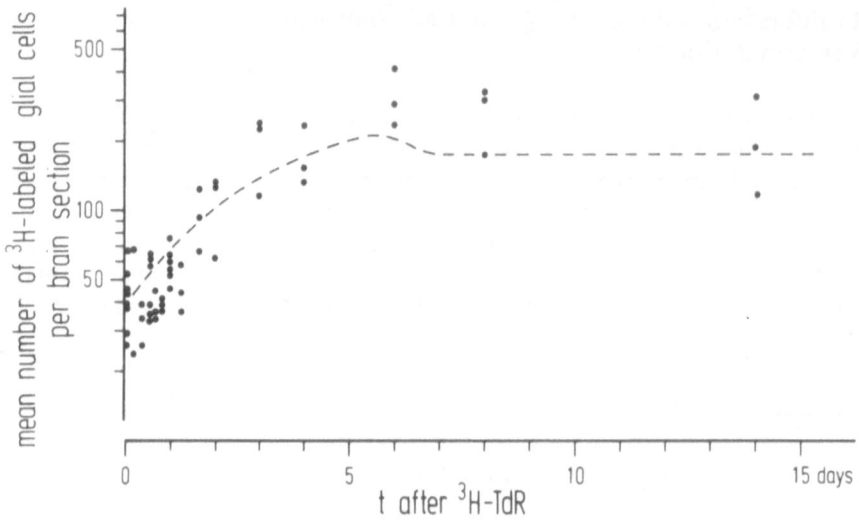

Fig. 13. Mean number of ³H-labeled glial cells per brain section (adult mouse) as a function of time after ³H-TdR injection. *Dots*, experimental values for individual animals; *broken line*, theoretical curve (see text)

interval in which probably no labeled glial cells have yet completed mitosis, and thus the number of labeled cells remains constant.

The experimental values in Fig. 13 were compared with a theoretical curve (broken line). This curve proceeds from the data measured at the start of the experiment and takes into account the following factors: (1) the cell cycle parameters of the glial cells; (2) the 17% loss of the daughter cells after each mitosis, as determined in Sect. 5.1.1.1; (3) the fact that glial cells are scored as labeled when a certain grain count per nucleus is reached (a circumstance which must be considered when comparing autoradiographs with different exposure times); and (4) the assumption of steady state growth. As is seen, the experimental values are in good agreement with the theoretical assumption.

ii) Figure 14 gives the measured mean grain counts per nucleus for the glial cells (dots) as well as for cells of the subependymal layer (diamonds, to be discussed later) on a semilog scale as a function of time after ³H-TdR injection. As the solid curve shows (this curve describes the course of experimental values), the mean grain counts first exhibit a steady decline for the first few days of the experiment. Then, starting about day 6, the curve flattens out and runs nearly parallel to the time axis.

Figure 14 also contains a dotted line which represents the theoretically expected decline in the mean grain count per nucleus for a steady state growth. As Fig. 14 shows, this curve does not coincide with the experimental data: While a mean grain count of 13 silver grains per nucleus was determined for the experimental curve in the interval from day 6 to day 14, the theoretical curve indicates a value of 37 grains per nucleus, or about 3 times the experimental value.

iii) A similar discrepancy between the experimental values and theoretical expectation arose on analysis of the frequency distribution of the grain count per nucleus. This analysis was done by correcting the relative frequency distributions of the grain count per nucleus obtained by the evaluation of autoradiographs (Fig. 15) for a uniform exposure time of 1 day and plotting them as cumulative frequency distributions [F(n) = P

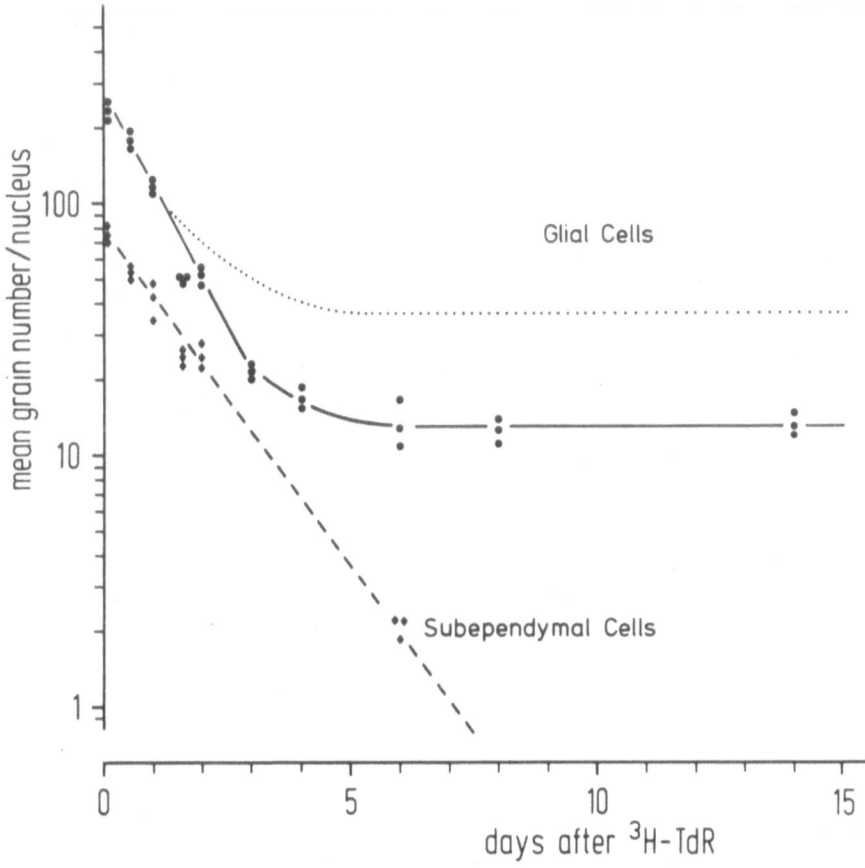

Fig. 14. Mean grain count per nucleus for labeled glial cells and cells of the subependymal layer in the brain of the adult mouse as a function of time after ³H-TdR injection. Symbols: experimental values for individual animals (*dots*, glial cells, *diamonds*, cells of subependymal layer). The mean grain counts per nucleus 1–48 h after ³H-TdR injection were multiplied by the factor 5 to correct for differences in exposure time. *Solid and broken lines*, experimental curve; *dotted line*, theoretical curve for glial cells (see text)

($\underline{n} \geqslant n$)]. Thus, in each case a cumulative distribution curve was constructed which was based on the frequencies of the cells with the highest grain counts per nucleus. Figure 16 shows three such cumulative frequency distribution curves plotted in semilog coordinates: Curve A shows the distribution of the grain count per nucleus 1 h after ³H-TdR injection, and curve B the distribution 14 days after ³H-TdR injection. A theoretical curve C is also shown, which is derived from the experimentally determined curve A (1 h after ³H-TdR) under certain assumptions. Curve C should arise from curve A after 14 days if a constant percentage of the daughter cells fails to proliferate further after each completed mitosis. Since all cells proliferating at the start of the experiment have already divided so many times during the 14-day period studied that their label has been diluted below detectable limits, curve C (like curve B, which is to be compared with it) is comprised of the sum of the labeled glial cells which have failed to proliferate further after the individual mitotic stages.

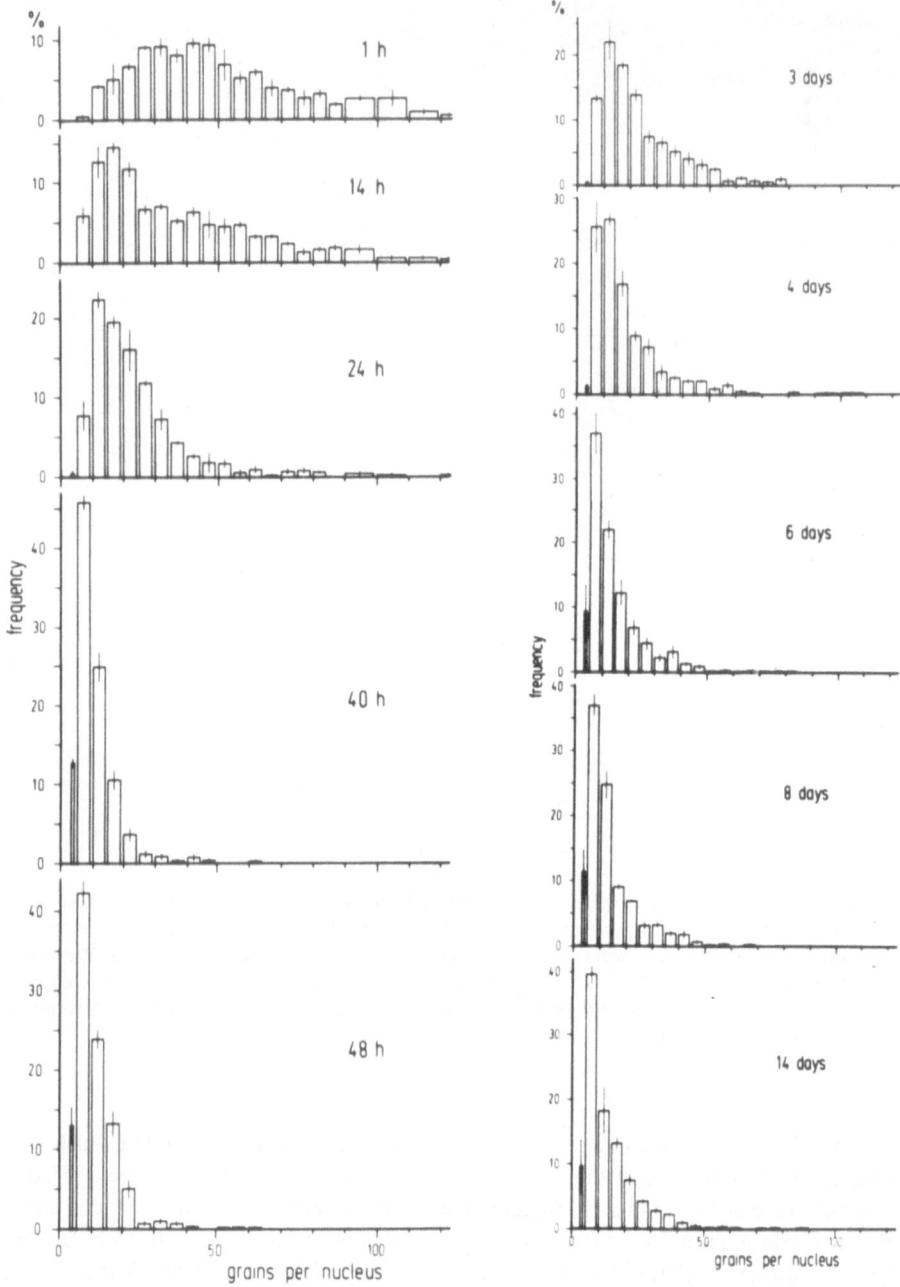

Fig. 15. Relative frequency distributions of grain count per nucleus for labeled glial cells in the brain of the adult mouse after injection of 500 μCi ^{3}H-TdR. *Left*, experiments 1–48 h after ^{3}H-TdR injection, exposure time 4 days. *Right*, experiments 3–14 days after ^{3}H-TdR injection; exposure time 20 days. The mean value of three individual distributions is given for each grain-count interval. The *vertical lines* indicate the standard error of the mean value

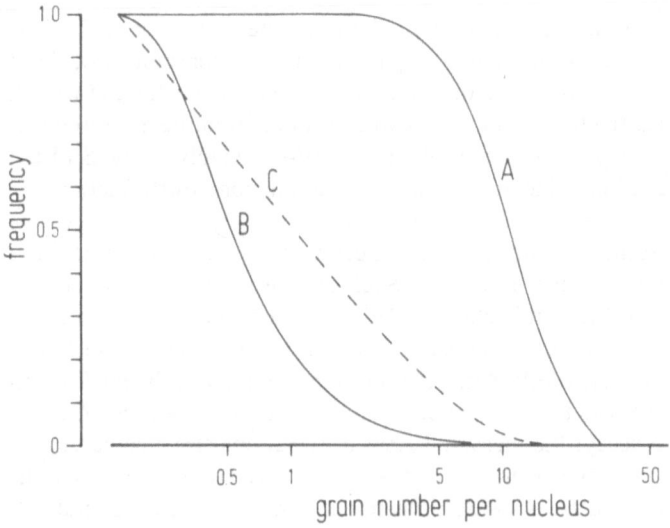

Fig. 16. Cumulative frequency distribution [F (n) = P (n ⩾ n)] of the grain count per nucleus for labeled glial cells in the brain of the adult mouse after ^{3}H-TdR injection. *A*, 1 h after ^{3}H-TdR injection, *B*, 14 days after ^{3}H-TdR injection; *C*, theoretical curve (see text). All three curves pertain to an exposure time of 1 day

As in the case of the mean grain count per nucleus (Fig. 14), a comparison of experimental curve B with theoretical curve C in Fig. 16 shows an apparent lack of agreement. Although glial cells with highly diverse labeling intensities were observed 14 days after the single injection of ^{3}H-TdR, marked differences were found with respect to quantitative composition. Fewer heavily and moderately heavily labeled glial cells were observed than had been expected in theory.

To explain the discrepancies found in ii and iii, the following factors were analyzed more closely:

1. The problem of correcting the results of autoradiographic experiments for a uniform exposure time (see Korr 1978a).

2. The possible reutilization of ^{3}H-DNA precursors (see Schultze 1969). In this case the glial cells would be labeled not only by the single ^{3}H-TdR injection, but also later on as the proliferating cells absorb ^{3}H-DNA fragments or metabolites contained in the blood.

3. The possible immigration of weakly labeled hematogenous cells (cf. Roessmann and Friede 1968; Oehmichen et al. 1973) or cells of the subependymal layer (cf. Lewis 1968c; Privat and Leblond 1972; Paterson et al. 1973). Such cells could not be distinguished histologically from the other labeled glial cells.

4. The damage from radiation; β-radiation from the ^{3}H-β-particles could cause a loss of heavily labeled glial cells in particular.

As a detailed discussion has shown (cf. Korr 1978a), there has been no experimental evidence to indicate that the factors listed above are actually responsible for the discrepancies observed.

On the other hand, it must be considered that the contradictions may rest upon faulty assumptions for constructing the theoretical curves, i.e., that factors other than the assumptions on cell cycle parameters and mode of proliferation are involved. For

48

example, it has always been assumed that the volume of the nucleus as well as the geometry of the nuclear surface of labeled neuroglial cells remain constant throughout the duration of the study. Thus, these factors are thought to remain unchanged when a cell passes from the growth fraction to the nongrowth fraction. The initial measurement 1 h after ^{3}H-TdR injection pertained to proliferating cells primarily in the S phase, while the measurement at 14 days included diploid cells in the nongrowth fraction.

In the still hypothetical case that these two factors do change, the result would be as follows: First let us assume that the passage of a cell into the nongrowth fraction is associated with an increase in its nuclear volume. Such a process has been described for human fibroblasts in vitro (Mitsui and Schneider 1976). An increase in nuclear volume would in turn lead to a dilution of the radioactive label and thus to lower grain counts per nucleus (cf. Appleton et al. 1969). With regard to the geometry of the nuclear surface, let us assume that besides the incorporated radioactivity and the Poisson distribution of the radioactive decay, the measured grain distributions also include the frequency distribution of the sectional surface of the nucleus as a factor. Changes in the distribution of the sectional surfaces would influence the steepness of the cumulative distribution curves in Fig. 16.

No experimental data are yet available on the changes that occur in the nuclear volume and nuclear surface geometry of neuroglial cells in the adult mouse brain. However, it is entirely conceivable that the aforementioned discrepancies regarding the decline in the mean grain count per nucleus (Fig. 14) and the grain distribution (Fig. 16) can be resolved when more is learned about these factors.

In summary, it can be said that while quantitative contradictions still exist in the assumption of steady state growth, it is nonetheless clear that neuroglial cells passed from the growth fraction to the nongrowth fraction during the 14-day period studied. It is also apparent that the proper quantitative evaluation of autoradiographic studies requires more data than have traditionally been obtained.

b) Passage of glial cells from the nongrowth fraction to the growth fraction. The passage of glial cells from the growth fraction to the nongrowth fraction was discussed in the preceding chapter. In the present section we shall examine the question of whether the opposite also occurs, i.e., whether cells of the nongrowth fraction again start to proliferate, and thus enter the growth fraction. Such a phenomenon is already well known in such fields as tumor cell biology (cf. Steel and Lamerton 1968; Epifanova and Terskikh 1969).

To demonstrate experimentally the entry of glial cells into the growth fraction, the following experiments (shown schematically in Fig. 17) were performed on adult

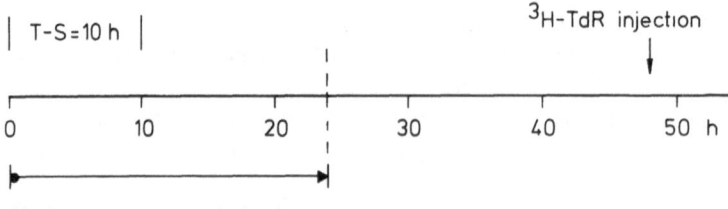

Fig. 17. Scheme of experiments for demonstrating the passage of cells from the nongrowth fraction into the growth fraction

mice: first; all proliferating glial cells were labeled by continuous infusion of ^{14}C-TdR. With an S phase length of about 10 h and a cycle time of 20 h, all proliferating cells are in the S phase once during a period corresponding to the difference between the cycle time and S phase length, i.e., 10 h. Thus, all proliferating cells can be labeled by ^{14}C-TdR infusion over this period. However, this is true only if the lengths of the cycle phases are constant. In order to label glial cells which may have a longer cycle time, the period during which ^{14}C-TdR was administered was extended far beyond the theoretically adequate period of 10 h: over 18 h in one test series (A) and over 24 h in a second test series (B). The ^{14}C-labeled TdR was administered to the mice by multiple injections at 6-h intervals. With an S phase length of 10 h, this is sufficient to label at least once all cells which pass through the S phase in the interval between the first (t = 0 h) and last ^{14}C-TdR injection.

Thirty h (experiment A) or 24 h (experiment B) after continous labeling with ^{14}C-TdR was stopped, the experimental animals received a single injection of ^{3}H-TdR and were killed 1 h later. One would expect that most of the labeled glial cells would proliferate further by the end of the experiment, although a portion of the cells will cease to proliferate. Now if only those cells which are prelabeled with ^{14}C are proliferating at the time of the ^{3}H-TdR injection, the autoradiographs should show purely ^{14}C-labeled glial cells as well as double-labeled glial cells, but no purely ^{3}H-labeled glial cells. When the two emulsion layer autoradiographs were evaluated, however, purely ^{3}H-labeled glial cells were found in addition to purely ^{14}C-and double-labeled cells (for details see Korr 1978a; Korr et al., to be published d).

The observation of double-labeled glial cells indicates that cells indeed proliferated further after ^{14}C labeling and were again in the S phase at the time of ^{3}H-TdR injection. The purely ^{3}H-labeled cells are cells which were in the S phase at the time of ^{3}H-TdR injection, but not during the period of multiple ^{14}C-TdR injections. There are two possible explanations for this:

1. The purely ^{3}H-labeled glial cells are proliferating cells which were outside the S phase (i.e., in phases G_2, M, and G_1) throughout the period in which ^{14}C-TdR was administered. If this assumption is correct, the purely ^{3}H-labeled glial cells would exhibit a cycle time far greater than 20 h. Based on considerations in which the time-course of the experiments was correlated with the phase age of the proliferating glial cells, a T value > 48 h could be estimated for these glial cells. It is true that no precise data are yet available on the range over which T_C can vary in glial cells; however, it can be assumed from the results of double labeling experiments using the labeled S phase method (cf. Sect. 4.1.4.3) that the adult mouse brain does not contain many glial cells which have a cycle time greater than 48 h. This suggests that the above interpretation is incorrect.

2. The second explanation for the occurrence of purely ^{3}H-labeled glial cells is that these cells were not proliferating at all during the time of the ^{14}C-TdR injections, i.e., were still part of the nongrowth fraction. This would mean that the purely ^{3}H-labeled glial cells did not start to proliferate (i.e., enter the growth fraction) until after the last ^{14}C-TdR injection.

This second interpretation is not contradicted by any other experimental results. It thus appears justified to interpret the observation of purely ^{3}H-labeled glial cells as experimental evidence that a constant passage of glial cells from the nongrowth fraction does indeed occur in the adult mouse brain.

c) Scheme of proliferation of neuroglia in the adult mouse brain. According to the experimental results discussed in the preceding sections, it appears that the proliferation of neuroglia in the adult mouse brain is accompanied by a constant exchange of cells between the growth fraction and nongrowth fraction. Accordingly, the pool of proliferating glial cells is continuously acquiring new members and losing old ones. This scheme of proliferation, together with the cell loss discussed before, is shown diagramatically in Fig. 18. According to this scheme the mode of neuroglia proliferation is as follows: nonproliferating glial cells enter the DNA synthesis phase, pass through one or more cycles with cell loss, and then reenter the nonproliferating pool. The observation that the ratio of ^{14}C-labeled astrocytes to ^{14}C-labeled oligodendrocytes remained nearly constant during the 14-day period studied (Korr 1978a) suggests that the proliferation scheme shown in Fig. 18 applies not only to the sum of the astrocytes and oligodendrocytes, but possibly to each of these cell types individually.

5.2.1.2 Studies on Glial Cells in the Young Rat Brain

All the experiments described in Sect. 5.2.1.1a and 5.2.1.1b were also performed in the same manner on 14-day-old rats. These results will not be discussed in detail (for this see Korr 1978a; Korr et al., to be published c). It is sufficient here to point out that the results were in complete agreement with the results given earlier for the adult mouse: Glial cells continuously leave the proliferating pool in the four investigated regions of the forebrain in the young rat, while other glial cells enter it. It has not yet been determined whether differences exist between astrocytes and oligodendrocytes with regard to these exchange processes.

5.2.1.3 Studies on Glial Cells and Their Precursors in the Prenatal Rat and Mouse Brain

Autoradiographic studies along the lines of those described for adult mice and young rats have not previously been done on rodents in the prenatal phase of life. Nevertheless, it can be concluded from the appearance of labeled neuroglial cells in the brain of young animals injected with ^{3}H-TdR at specific points during embryonic development (Sect. 3.2.2) that glial cells leave the growth fraction early in prenatal life. As mentioned

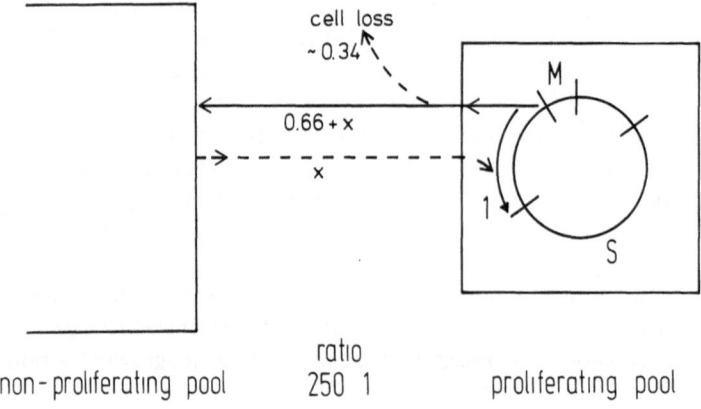

Fig. 18. Scheme of glial cell proliferation in the brain of the adult mouse

earlier, continuous divisions of the prenatally labeled cells would have led to a strong dilution of the label.

As experiments by Hinds (1968) on the mouse olfactory bulb have shown, non-proliferating neuroglial cells (probably astrocytes) can be observed from the 11−12th embryonic day on. Assuming that the histologic differentiation of Berry and Rogers (1966) with respect to astrocytes, oligodendrocytes, and microglial cells is correct (cf. Sect. 3.2.2), nonproliferating oligodendrocytes and astrocytes or their precursors are probably present from about E 18 on.

Biesold at al. (1976) offer an idea of the degree to which neuroglial cells leave the growth fraction during pre- and early postnatal ontogeny: These authors injected ^{3}H-TdR between E 14 and P 21 and killed the animals between P 22 and P 27. They then determined the percentages of labeled cells for "ectodermal nonneuronal cells" in the dorsal lateral geniculate nucleus. According to these studies, the cells do not leave the growth fraction uniformly; distinct maxima (around P 5 and P 15) and minima (around E 20 and P 11) are evident.

There is as yet no evidence that nonproliferating glial cells later reenter the growth fraction, resulting in an exchange between the growth fraction and nongrowth fraction in the prenatal brain.

5.2.2 Endothelial Cells

The experiments described in Sect. 5.2.1.1 yielded essentially the same results for the endothelial cells in the forebrain of the adult mouse and 14-day-old rat as for the neuroglial cells; i.e., these cells also exhibit a permanent exchange between the growth fraction and nongrowth fraction (for details see Korr 1978a; Korr et al., to be published a). Thus, the proliferation scheme shown in Fig. 18 appears also to apply to endothelial cells during postnatal ontogeny. It is still unclear, however, whether differences exist between these cells and the neuroglia from a quantitative standpoint.

From the autoradiographic results presented in Sect. 3.2.6, it can be concluded that, as in the case of the neuroglia, in the case of endothelial cells in the prenatal brain also a nongrowth fraction exists from about E 15 on.

5.2.3 Cells of the Subependymal Layer

As experiments following a single injection of ^{3}H-TdR and a survival time of 1 h to 14 days have shown (see Sect. 5.2.1.1a), the mean grain count per nucleus of the subependymal cells in the adult mouse brain declines continuously over a period of several days (broken curve in Fig. 14). Fourteen days after ^{3}H-TdR injection, labeled subependymal cells were no longer found in the brain regions investigated, although the autoradiograph exposure time was 20 days. Smart (1961) also found that labeled subependymal cells could no longer be observed in the adult mouse brain 10 days after injection of 10 μCi ^{3}H-TdR per gram body weight and an exposure time of 28 days. As both these studies show, the labeled subependymal cells appear to have divided continuously with a cycle time of 18−24 h, at least during the 14-day period studied. The changes in the grain-distribution histograms (Fig. 19; see also Fig. 2) also support this assumption: the percentage of cells with very low grain counts (1−2 grains per nucleus) increases continuously with time after ^{3}H-TdR injection; that means, the range of the distribution

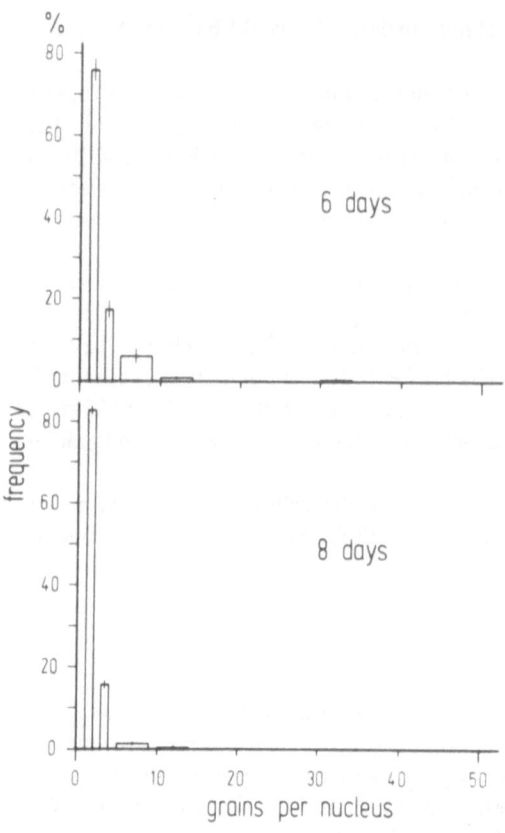

Fig. 19. Relative frequency distributions of grain count per nucleus for labeled cells of the subependymal layer in the brain of the adult mouse after injection of 500 μCi ^{3}H-TdR. Exposure time: 20 days. The mean value of three individual distributions is given for each grain-count interval. The *vertical lines* indicate the standard error of the mean value

grows smaller. Thus, it appears that at 6 and 8 days, for example, there are no longer any labeled cells present which entered the nongrowth fraction considerably earlier (e.g., 1 or 2 days after ^{3}H-TdR injection).

If the labeled, and thus all proliferating subependymal cells divide continuously, this behavior should correlate with the measured percentages of labeled cells as a function of time after a single ^{3}H-TdR injection. To test this assumption, the experimental values were compared with values predicted by theory. In Fig. 20 the solid curve represents the measured percentage of labeled subependymal cells in the adult mouse brain as a function of time after ^{3}H-TdR injection (dots: values for individual experimental animals). As Fig. 20 shows, the percentage of labeled cells doubles initially and then remains more or less constant until 48 h after injection. Six and 8 days after injection of ^{3}H-TdR, higher experimental values are observed than in the interval from 24 to 48 h. However, due to the presence of many very weakly labeled cells (cf. Fig. 19), the measured percentage of labeled cells at 6 and 8 days after ^{3}H-TdR injection may well have been too low.

The broken curve in Fig. 20 was derived on the basis of theoretical considerations (for details on its construction see Korr 1978a). This curve represents the percentage of labeled subependymal cells as a function of time after ^{3}H-TdR injection, under the assumption that the proliferating cells divide continuously. The curve originates from the value measured at 1 h after ^{3}H-TdR injection and takes into account data published elsewhere on the proliferation of the subependymal cells: specifically that the ratio of

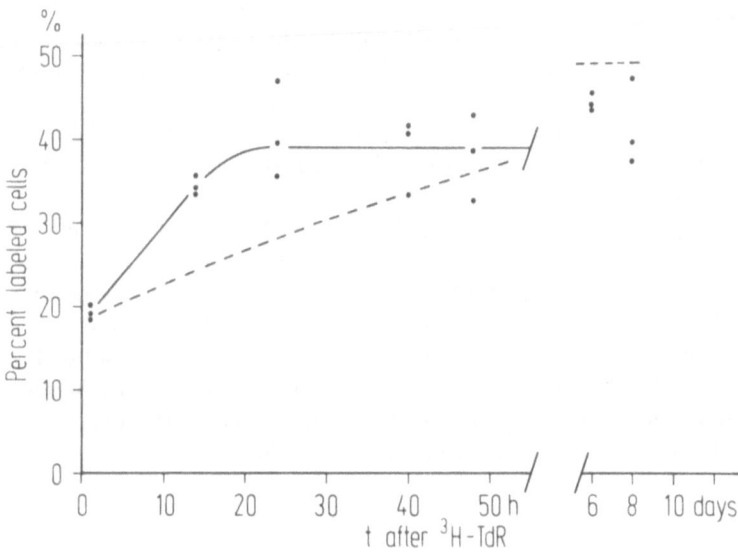

Fig. 20. Percentage of labeled subependymal cells in the brain of the adult mouse as a function of time after ^{3}H-TdR injection. *Dots and solid line,* experimental values for individual animals, *broken line,* theoretical curve (see text)

proliferating cells in the S phase to all proliferating cells (N_S/N_T) is equal to 0.5 (see Table 1) and that the cell loss per completed mitosis amounts to 9% of all daughter cells. The theoretical curve is further based on a value of 20 h for the cell cycle time and the assumption that the labeled cells are not limited to the range of the S phase, but are distributed over the entire cell cycle after a completed mitosis. The latter assumption has no bearing on further conclusions but simplifies construction of the theoretical curve.

A comparison of the two curves in Fig. 20 shows that the experimental values considerably exceed the theoretical values in the range up to about 40 h after ^{3}H-TdR injection. After longer periods (48 h, but particularly 6 and 8 days), the experimental values largely agree with the theoretical expectation.

In interpreting the curves in Fig. 20, it must first be remembered that the value measured at 1 h after ^{3}H-TdR injection is obtained from the ratio of the proliferating cells in S to all proliferating and nonproliferating cells. If the experimental values in the range up to 40 h prove to be higher than the theoretical ones under the assumption of continuous cell division, this means that unlabeled cells have already left the subependymal layer during this period. The fact that cells really do leave the subependymal layer is indicated by the observation that although the cells of the subependymal layer continue to proliferate for life, the number of cells in the layer does not increase (cf. Hopewell 1971).

If unlabeled cells leave the subependymal layer in the period up to 40 h after ^{3}H-TdR injection, the question arises whether these cells are in the growth fraction or the nongrowth fraction. Now it is assumed that if unlabeled proliferating cells leave the subependymal layer, labeled proliferating cells will also leave it in the same manner. Due to the constancy of the number of subependymal cells, however, the loss of proliferating cells would result in little or no increase in the percentage of labeled subependymal cells with time. This conclusion is in conflict with experimental data. It appears, then, that at least the majority of cells that leave the subependymal layer are cells from the nongrowth fraction. This means further that due to the constant size of the growth fraction and nongrowth fraction, the pool of nonproliferating cells is always filled; in other words, the passage of cells from the growth fraction to the nongrowth fraction must take place within the subependymal layer.

In summary, the analysis of autoradiographic findings after a single injection of ^{3}H-TdR indicates that a portion of the cells of the subependymal layer are continuously dividing and can thus be regarded as a germinal layer (cf. also Privat 1975). Approximately 9% of the daughter cells become pyknotic. Others enter the nongrowth fraction and leave the subependymal layer. It can be concluded from the grain-distribution histograms that the cells which cease proliferating remain only a short time in the subependymal layer. It is unlikely, therefore, that nonproliferating cells in the subependymal layer will have an opportunity to reenter the growth fraction.

5.2.4 Neurons

From the results of Schultze et al. (1974a,b) on the time at which the neuroepithelial cells start to proliferate in specified brain areas conclusions can be drawn on the genesis of the individual nuclei of the brain and, more generally, on the mode of proliferation of the precursor cells of the neurons. As the example of the four nuclei investigated by the aforementioned authors shows (supraoptic nucleus, n. ruber, n. cochlearis ventralis, Purkinje cells of the cerebellum), the associated precursor cells start to proliferate at approximately the same time. Their number is increased exponentially by continuous cell divisions. The end of proliferation, which varies from one brain nucleus to the next, appears in this case to be correlated with the number of cells. If this example can be generalized, it appears that the individual nuclei of the brain are formed in the shortest possible time by continuous divisions of their precursor cells.

This should not be taken to mean that the development of the neuroepithelial cells is predetermined from the start of their proliferation with regard to a particular type of neuron. It is much more reasonable to assume that the neuroepithelial cells first proliferate homogenously, accompanied by a constant interchange between the different "family trees" of the later neuron types.

If one considers the neuroepithelial cells which later become neurons in their entirety, rather than by different brain nuclei, their mode of proliferation is such that cells constantly pass into the nongrowth fraction from early embryonic development on (e.g., from about E 10 on for the mitral cells of the mouse olfactory bulb: Hinds 1968). The same proliferative behavior has been demonstrated for neuroepithelial cells which later develop into neuroglial cells.

From the example of the neuroblasts, it is clear that there is no way for the corresponding neuroepithelial cells to reenter the growth fraction from the nongrowth fraction, and one cannot speak of an exchange between these two fractions.

5.2.5 Ependymal Cells and Epithelial Cells of the Choroid Plexus

Heavily labeled cells were found in the brain of 25-day-old rats following prenatal injection of ^{3}H-TdR after E 11 (ependymal cells of the fourth ventricle) or after E 11/12 (epithelial cells of the choroid plexus in the fourth ventricle) (Korr 1978a). This shows that, as in the case of the astrocytes and certain other neuron precursors, cells leave the growth fraction starting at a very early stage of embryonic development. It is not known whether some of these nonproliferating cells can later reenter the growth fraction, resulting in an exchange between the growth fraction and nongrowth fraction.

5.2.6 Microglia and Pericytes

As indicated by the results of autoradiographic studies presented in Sect. 3.2.7 the microglia and/or pericytes exhibit a continuous passage of proliferating cells into the nongrowth fraction from about E 18 to the end of the proliferation phase (at about the 3rd week of life). Whether the reverse process also occurs (i.e., the passage of cells from the nongrowth fraction to the growth fraction) is unknown.

5.3 Comparison of the Mode of Proliferation of Various Cell Types During Pre- and Postnatal Ontogeny

The mode of proliferation described here for the various cell types in the CNS of the rat and mouse is summarized in Fig. 21. Fig. 21 presents a schematized overview of the pre- and postnatal phases of cell proliferation (see Sect. 3.2) and of the mode of proliferation (Sect. 5) of the different cell types.

The period of proliferation is represented by black bars, whose thickness gives an approximate idea of the degree of proliferation. The symbols below the bars in Fig. 21 refer to the mode of proliferation. In all the cell types, cells cease to proliferate both prenatally and postnatally (see •). The opposite phenomenon, i.e., that nonproliferating cells again start to proliferate, has so far been demonstrated experimentally only for neuroglia and endothelial cells in postnatal life (see ▲). In these cell types, therefore, there is a constant exchange between the growth fraction and nongrowth fraction. Moreover, the neuroglial cells, endothelial cells, the cells of the subependymal layer, and the precursor cells of the neurons proliferate with cell loss (see x). This has been demonstrated experimentally for the postnatal proliferation phase. This mode of proliferation may also be present in prenatal ontogeny, however.

6 Summary

The present work deals with the proliferation of various cell types in the brain of the untreated rat and mouse during pre- and postnatal ontogeny; specifically, the neuron precursor cells, neuroglial cells, cells of the subependymal layer, ependymal cells, epithelial cells of the choroid plexus, endothelial cells, as well as microglia and pericytes. Discussions center on our own autoradiographic studies with ^{3}H- and/or ^{14}C-thymidine.

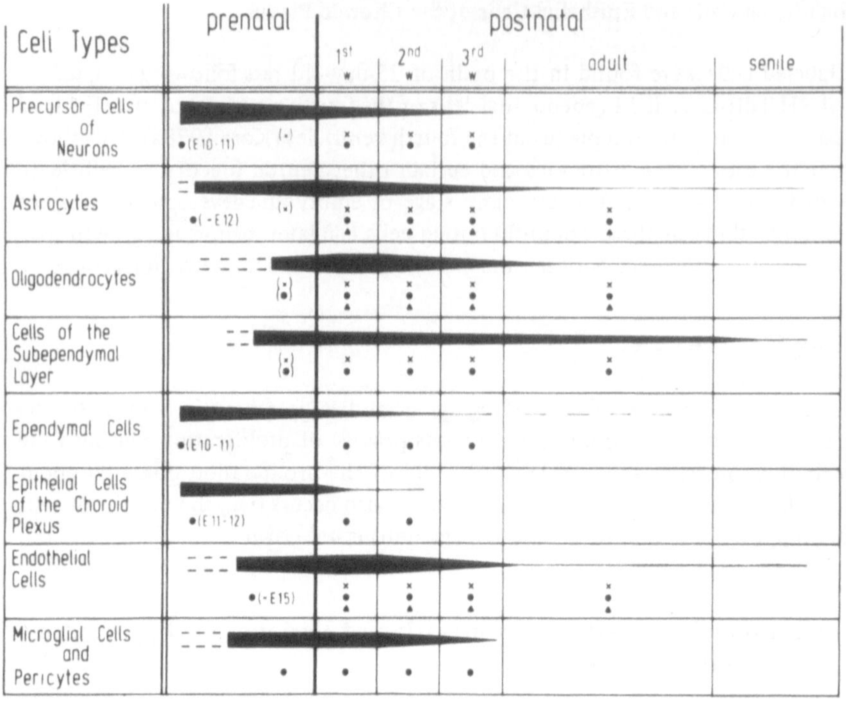

Cell Types	prenatal	postnatal				
		1st w	2nd w	3rd w	adult	senile
Precursor Cells of Neurons	●(E 10-11) (•)	× •	× •	× •		
Astrocytes	●(-E 12) (•)	× • ▲	× • ▲	× • ▲	× • ▲	
Oligodendrocytes	(:•)	× • ▲	× • ▲	× • ▲	× • ▲	
Cells of the Subependymal Layer	(:•)	× •	× •	× •	× •	
Ependymal Cells	●(E 10-11)	•	•	•		
Epithelial Cells of the Choroid Plexus	●(E 11-12)	•	•			
Endothelial Cells	●(-E15)	× • ▲	× • ▲	× • ▲	× • ▲	
Microglial Cells and Pericytes	•	•	•	•		

▬ Period of proliferation
× Proliferation with cell loss
• Passage of cells from the growth fraction to the non growth fraction
▲ Passage of cells from the non growth fraction into the growth fraction

Fig. 21. Diagram of the proliferation period and mode of proliferation of various cell types in the brain of untreated rats and mice during pre- and postnatal ontogeny

For adult animals in particular, little or no information has been published to date on the proliferation of the different cell types, because most methods of investigating cell kinetics are unsuitable for such applications. For this reason, the special autoradiographic techniques employed in each case are discussed in some detail. The results obtained are presented in conjunction with data from the literature, and conclusions are drawn with respect to the following three points:

1. Period of proliferation. The cell types studied can be divided into two groups according to the length of their proliferation phase: (a) The precursor cells of neurons, the ependymal cells, the epithelial cells of the choroid plexus, as well as the microglia and pericytes proliferate mainly prenatally, as well as for a short period of about 3-4 weeks after birth. With the exception of a very few ependymal cells, these cell types no longer proliferate in the brain of untreated adult animals. (b) Astrocytes, oligodendrocytes, the cells of the subependymal layer, and endothelial cells proliferate throughout postnatal life, even in the brain of very old animals. The number of proliferating cells in adult animals is very small, however.

2. Cell cycle parameters. The cell cycle time for all cell types is very short (about 7–8 h) during the early embryonic period. The S phase length is about 5 h. These times lengthen as fetal age increases: Toward the end of the fetal period all proliferating cell types have more or less consistent cycle times of about 18–20 h and S phase lengths of about 6–11 h. These relatively short phase lengths persist into postnatal life: In the brain of adult mice, for example, the astrocytes, oligodendrocytes, subependymal cells, and endothelial cells have cycle times of about 20 h and S phase lengths of about 10 h. In the case of neuroglial cells and endothelial cells, the percentage of proliferating cells declines sharply with age.

3. Mode of proliferation. Neuroglia, cells of the subependymal layer, and endothelial cells proliferate with a constant cell loss in the adult animal brain. This is probably also characteristic of the precursor cells of neurons which proliferate both prenatally and for the first weeks after birth. At the same time, the neuroglia and endothelial cells proliferate postnatally with a continuous exchange of cells between the growth fraction and the nongrowth fraction. In this process proliferating cells leave the growth fraction after each mitosis. Some of these cells are then replaced by cells which were previously in the nongrowth fraction. The departure of cells from the growth fraction was observed for all investigated cell types in both pre- and postnatal ontogeny.

References

Adelstein SJ, Lyman CP, O'Brien RC (1964) Variations in the incorporation of thymidine into the DNA of some rodent species. Comp Biochem Physiol 12: 223–231

Adrian EK Jr (1971) The metabolic stability of nuclear DNA. In: Cameron IL, Trasher JD (eds) Cellular and molecular renewal in the mammalian body. Academic Press, New York London, pp 25–44

Adrian EK Jr, Walker BE (1962) Incorporation of thymidine-^{3}H by cells in normal and injured mouse spinal cord. J Neuropathol Exp Neurol 21: 597–609

Adrian EK Jr, Williams MG (1973) Cell proliferation in injured spinal cord. An electron microscopic study. J Comp Neurol 151: 1–24

Adrian EK Jr, Williams MG, George FC (1978) Fine structure of reactive cells in injured nervous tissue labeled with ^{3}H-thymidine injected before injury. J Comp Neurol 180:815–840

Aherne WA, Camplejohn RS, Wright NA (1977) An introduction to cell population kinetics. Arnold, London

Alpen EL, Cranmore D (1959) Observations on the regulation of erythropoiesis and on cellular dynamics by Fe^{59} autoradiography. In: Stohlman F Jr (ed) The kinetics of cellular proliferation. Grune & Stratton, New York London, pp 290–300

Altman J (1962a) Are new neurons formed in the brains of adult mammals? Science 135: 1127–1128

Altman J (1962b) Autoradiographic study of degenerative and regenerative proliferation of neuroglia cells with tritiated thymidine. Exp Neuol 5: 302–318

Altman J (1963) Autoradiographic investigation of cell proliferation in the brains of rats and cats. Anat Rec 145: 573–591

Altman J (1966) Autoradiographic and histological studies of postnatal neurogenesis. II. A longitudinal investigation of the kinetics, migration and transformation of cells incorporating tritiated thymidine in infant rats, with special reference to postnatal neurogenesis in some brain regions. J Comp Neurol 128: 431–474

Altman J (1967) Postnatal growth and differentiation of the mammalian brain, with implications for a morphological theory of memory. In: Quarton G, Melnechuk T, Schmitt FO (eds) The neurosciences. A study program. Rockefeller University Press, New York, pp 723–743

Altman J (1969a) DNA metabolism and cell proliferation. In: Lajtha A (ed) Structural neurochemistry, Plenum Press, New York London (Handbook of neurochemistry, Vol II, pp 137–182)

Altman J (1969b) Autoradiographic and histological studies of postnatal neurogenesis. IV. Cell proliferation and migration in the anterior forebrain, with special reference to persisting neurogenesis in the olfactory bulb. J Comp Neurol 137: 433–458

Altman J (1970) Postnatal neurogenesis and the problem of neural plasticity. In: Himwich WA (ed) Developmental neurobiology. Thomas, Springfield, pp 197–237

Altman J, Bayer SA (1977) Time of origin and distribution of a new cell type in the rat cerebellar cortex. Exp Brain Res 29: 265–274

Altman J, Bayer SA (1978a) Prenatal development of the cerebellar system in the rat. I. Cytogenesis and histogenesis of the deep nuclei and the cortex of the cerebellum. J Comp Neurol 179: 23–48

Altman J, Bayer SA (1978b) Prenatal development of the cerebellar system in the rat. II. Cytogenesis and histogenesis of the inferior olive, pontine gray, and the precerebellar reticular nuclei. J Comp Neurol 179:49–76

Altman J, Bayer SA (1978c) Development of the diencephalon in the rat. I. Autoradiographic study of the time of origin and settling patterns of neurons of the hypothalamus. J Comp Neurol 182: 945–972

Altman J, Bayer SA (1978d) Development of the diencephalon in the rat. III. Ontogeny of the specialized ventricular linings of the hypothalamic third ventricle. J Comp Neurol 182: 995–1016

Altman J, Das GD (1965) Postnatal origin of microneurones in the rat brain. Nature 207: 953–956

Altman J, Nicholson JL (1971) Cell pyknosis in the cerebellar cortex of infant rats following low-level X-irradiation. Radiat Res 46:476–489

Anderson CH (1978) Time of neuron origin in the anterior hypothalamus of the rat. Brain Res 154: 119–122

Angevine JB Jr (1970) Critical cellular events in the shaping of neural centers. In: Schmitt FO (ed) The neurosciences second study program. Rockefeller University Press, New York, pp 62–72

Appleton TC, Pelc SR, Tarbit MH (1969) Formation and loss of DNA in intestinal epithelium. J Cell Sci 5: 45–55

Asbury AK (1967) Schwann cell proliferation in developing mouse sciatic nerve. A radioautographic study. J Cell Biol 34: 735–743

Atlas M, Bond VP (1965) The cell generation cycle of the eleven-day mouse embryo. J Cell Biol 26: 19–24

Bär T, Wolff JR (1972) The formation of capillary basement membranes during internal vascularization of the rat's cerebral cortex. Z. Zellforsch 133:231–248

Baron M, Gallego A (1972) The relation of the microglia with the pericytes in the cat cerebral cortex. Z Zellforsch 128: 42–57

Baserga R, Tyler SA, Kisieleski WE (1963) The kinetics of growth of the Ehrlich tumor. A comparative study of the kinetics of cellular proliferation with the use of tritiated thymidine. Arch Pathol Lab Med 76: 9–13

Bayer SA (1977) Glial recovery patterns in rat corpus callosum after X irradiation during infancy. Exp Neurol 54: 209–216

Bayer SA (1979) The development of the septal region in the rat. I. Neurogenesis examined with ^{3}H-thymidine autoradiography. J Comp Neurol 183:89–106

Berry M, Rogers AW (1966) Histogenesis of mammalian neocortex. In: Hassler R, Stephan H (eds) Evolution of the forebrain. Phylogenesis and ontogenesis of the forebrain. Thieme, Stuttgart, pp 197–205

Biesold D, Brückner G, Mares V (1976) An autoradiographic study of gliogenesis in the rat lateral geniculate nucleus (LGN). Brain Res 104: 295–301

Blakemore WF (1972) Microglial reactions following thermal necrosis of the rat cortex: An electron microscope study. Acta Neuropathol (Berl) 21: 11–22

Blakemore WF (1975) The ultrastructure of normal and reactive microglia. Acta Neuropathol [Suppl] (Berl) 6:273–278

Boulder Committee (1970) Embryonic vertebrate central nervous system: Revised terminology. Anat Rec 166:257–262

Boya J (1976) An ultrastructural study of the relationship between pericytes and cerebral macrophages. Acta Anat (Basel) 95: 598–608

Boya J, Calvo J, Prado A (1979) The origin of microglial cells. J Anat 129:177–186

Bradley WG, Asbury AK (1970) Duration of synthesis phase in neurilemma cells in mouse sciatic nerve during degeneration. Exp Neurol 26: 275–282

Brichova H (1972) Contribution to the question of the existence and function of microglia cells in the rat CNS. Folia Morphol (Praha) 20:85–87

Brizzee KR, Vogt J, Kharetchko X (1964) Postnatal changes in glia/neuron index with a comparison of methods of cell enumeration in the white rat. Prog Brain Res 4:136–149

Brückner G, Mares V, Biesold D (1976) Neurogenesis in the visual system of the rat. An autoradiographic investigation. J Comp Neurol 166: 245–256

Bryant BJ (1966) The incorporation of tritium from thymidine into proteins of the mouse. J Cell Biol 29: 29–36

Burholt DR, Schultze B, Maurer W (1973) Autoradiographic confirmation of the mitotic division of every mouse jejunal crypt cell labeled with ^{3}H-thymidine. Evidence against the existence of cells synthesizing metabolic DNA. Cell Tissue Kinet 6: 229–237

Burns FJ, Tannock IF (1970) On the existence of a G_0-phase in the cell cycle. Cell Tissue Kinet 3: 321–334

Cameron IL (1971) Cell proliferation and renewal in the mammalian body. In: Cameron IL, Thrasher JD (eds.) Cellular and molecular renewal in the mammalian body. Academic Press, New York, London, pp 45–85

Cammermeyer J (1970) The life history of the microglial cell: A light microscopic study. Neurosci Res Program Bull 3:43–129

Cavanagh JB, Lewis PD (1969) Perfusion-fixation, colchicine and mitotic activity in the adult rat brain. J Anat 104: 341–350

60

Cerutti PA (1974) Excision repair of DNA base damage. Life Sci 15: 1567–1575

Chauhan AN, Lewis PD (1979) A quantitative study of cell proliferation in ependyma and choroid plexus in the postnatal rat brain. Neuropathol Appl Neurobiol 5: 303–309

Dalton MM, Hommes OR, Leblond CP (1968) Correlation of glial proliferation with age in the mouse brain. J Comp Neurol 134: 397–400

Das GD (1977) Gliogenesis during embryonic development in the rat. Experientia 33: 1648–1649

DeMaertelaer V, Galand P (1975) Some properties of a 'G$_o$' model of the cell cycle. I. Investigation on the possible existence of natural constraints on the theoretical model in steady-state conditions. Cell Tissue Kinet 8: 11–22

Denham S (1967) A cell proliferation study of the neural retina in the two-day rat. J Embryol Exp Morphol 18: 53–66

Dill RE, Walker BE (1966) Pineal cell proliferation in the mouse. Proc Soc Exp Biol Med 121: 911–912

Djordjevic B, Evans RG, Perez AG, Weill MK (1969) Spontaneous unscheduled DNA synthesis in G$_1$ HeLa cells. Nature 224: 803–804

Ellenberger C Jr, Hanaway J, Netsky MG (1969) Embryogenesis of the inferior olivary nucleus in the rat: A radioautographic study and a re-evaluation of the rhombic lip. J Comp Neurol 137: 71–88

Epifanova OI, Terskikh VV (1969) On the resting periods in the cell life cycle. Cell Tissue Kinet 2: 75–93

Evans RG, Djordjevic B, Perez AG (1970) Temperature dependence of thymidine uptake and stability of thymidine incorporated during unscheduled DNA synthesis in irradiated and unirradiated HeLa cells. Int J Radiat Biol 18: 277–280

Feinendegen LE (1967) Tritium-labeled molecules in biology and medicine. Academic Press, New York London

Fleischhauer K (1967) Postnatale Entwicklung der Neuroglia. Zentralbl Neurol Psychiatr 188: 385–386

Fleischhauer K (1972) Ependyma and subependymal layer. In: Bourne GH (ed) The structure and function of nervous tissue, vol 6. Academic Press, New York London, pp 1–46

Fried J (1968) Estimating the median generation time of proliferating cell systems in steady state. Biophys J 8: 710–729

Fried J (1970) Mean, geometric mean, or median grain count in cell cycle studies. Exp Cell Res 59: 447–451

Fujita S (1962) Kinetics of cell proliferation. Exp Cell Res 28: 52–60

Fujita S (1963) The matrix cell and cytogenesis in the developing central nervous system. J Comp Neurol 120: 37–42

Fujita S (1965) The matrix cell and histogenesis of the nervous system. Laval Med 36: 125–130

Fujita S (1966) Application of light and electron microscopic autoradiography to the study of cytogenesis of the forebrain. In: Hassler R, Stephan H (eds) Evolution of the forebrain. Phylogenesis and ontogenesis of the forebrain. Thieme, Stuttgart, pp 180–196

Fujita S (1967) Quantitative analysis of cell proliferation and differentiation in the cortex of the postnatal mouse cerebellum. J Cell Biol 32: 277–287

Fujita S, Kitamura T (1975) Origin of brain macrophages and the nature of the so-called microglia. Acta Neuropathol [Suppl] (Berl) 6: 291–296

Fujita S, Shimada M, Nakamura T (1966) H^3-thymidine autoradiographic studies on the cell proliferation and differentiation in the external and the internal granular layers of the mouse cerebellum. J Comp Neurol 128: 191–208

Gerhard H, Schultze B, Maurer W (1973) Quantitative Untersuchungen über Wachstum und Polyploidisierung bei der Regeneration der CCl$_4$-Leber der Maus. Virchows Archiv [Cell Pathol] 14: 345–359

Glücksmann A (1951) Cell death in normal vertebrate ontogeny. Biol Rev 26: 59–86

Goldspink DF, Goldberg AL (1973) Problems in the use of (Me-^{3}H) thymidine for the measurement of DNA synthesis. Biochim Biophys Acta 299: 521–532

Gracheva ND (1964) Kinetics and topography of the cell proliferation in the embryogenesis of the nervous system of white rats. In: Zhinkin LN, Zavarzin AA (eds) A study of cell cycles and metabolism of nuclei acids during differentiation of the cells. Nauka, Moskau Leningrad, pp 90–106

Gracheva ND (1969) Autoradiographic investigation of proliferative activity of the rat brain subependymal cells. Tsitologiya 11: 1521–1527

Haas RJ, Werner J, Fliedner TM (1970) Cytokinetics of neonatal brain cell development in rats as studied by the "complete ^{3}H-thymidine labeling method". J Anat 107:421–437

Hagemann E, Schmidt G (1960) Ratte und Maus. Versuchstiere in der Forschung. Gruyter, Berlin

Hager H (1968) Pathologie der Makro- und Mikroglia im elektronen-mikroskopischen Bild. Acta Neuropathol [Suppl] (Berl) 4 :86–97

Hager H (1975) EM findings on the source of reactive microglia on the mammalian brain. Acta Neuropathol [Suppl] (Berl) 6: 279–283

Hanawalt PC (1972) Repair of genetic material in living cells. Endeavour 31: 83–87

Hansson HA, Sourander P (1964) Studies on cultures of mammalian retina. Z Zellforsch 62: 26–47

Hicks SP, D'Amato CJ (1968) Cell migrations to the isocortex in the rat. Anat Rec 160: 619–634

Hilscher W, Maurer W (1962) Autoradiographische Bestimmung der Dauer der DNS-Verdoppelung und ihres zeitlichen Verlaufs bei Spermatogonien der Ratte durch Doppelmarkierung mit [14]C- und ^{3}H-Thymidin. Naturwissenschaften 49: 352

Hinds JW (1968) Autoradiographic study of histogenesis in the mouse olfactory bulb. I. Time of origin of neurons and neuroglia. J Comp Neurol 134:287–304

His W (1887) Die Entwicklung der ersten Nervenbahnen beim menschlichen Embryo. Übersichtliche Darstellung. Arch Anat Physiol Anat Abt (Leipzig) 92:368–378

His W (1889) Die Neuroblasten und deren Entstehung im embryonalen Rückenmark. Abh Math-Phys Klasse Königl Sächs Ges Wiss 15:331–372

Hommes OR. Leblond CP (1967) Mitotic divison of neuroglia in the normal adult rat. J Comp Neurol 129:269–278

Hopewell JW (1971) A quantitative study on the mitotic activity in the subependymal plate of adult rats. Cell Tissue Kinet 4:273–278

Hoshino K, Matsuzawa T, Murakami U (1973) Characteristics of the cell cycle of matrix cells in the mouse embryo during histogenesis of telencephalon. Exp. Cell Res 77:89–94

Howard A, Pelc SR (1949) P-32 autoradiographs of mouse testis. Heredity (Lond) 3: 383

Howard A, Pelc SR (1950) P-32 autoradiographs of the mouse testis. Preliminary observations of the timing of spermatogenic stages. Br J Radiol 23:634–641

Howard A, Pelc SR (1951) Nuclear incorporation of P-32 as demonstated by autoradiographs. Exp Cell Res 2:178–187

Howard-Flanders P (1968) DNA repair. Annu Rev Biochem 37:175–200

Hugosson R, Källen B, Nilsson O (1968) Neuroglia proliferation studied in tissue culture. Acta Neuropathol (Berl) 11:210–220

Imamoto K, Leblond CP (1977) Presence of labeled monocytes, macrophages and microglia in a stab wound of the brain following an injection of bone marrow cells labeled with ^{3}H-uridine into rats. J Comp Neurol 174:255–280

Imamoto K, Leblond CP (1978) Radioautographic investigation of gliogenesis in the corpus callosum of young rats. II. Origin of microglial cells. J Comp Neurol 180:139–164

Imamoto K, Paterson J (1974) Sequential labeling of subependymal cells and of the three types of oligodendrocytes but absence of microglia labeling in the corpus callosum following injections of ^{3}H-thymidine into young rats. Anat Rec 178:380–381

Imamoto K, Paterson JA, Leblond CP (1978) Radioautographic investigation of gliogenesis in the corpus callosum of young rats. I. Sequential changes in oligodendrocytes. J Comp Neurol 180:115–138

Jacobson M (1978) Developmental neurobiology, 2nd edn. Plenum Press, New York London

Jones EG (1970) On the mode of entry of blood vessels into the cerebral cortex. J Anat 106: 507–520

Kaplan MS, Hinds JW (1977) Neurogenesis in the adult rat: electron microscopic analysis of light radioautographs. Science 197:1092–1094

Kauffman SL (1966) An autoradiographic study of the generation cycle in the ten-day mouse embryo neural tube. Exp Cell Res 42:67–73

Kauffman SL (1968) Lengthening of the generation cycle during embryonic differentiation of the mouse neural tube. Exp Cell Res 49:420–424

Kerns JM, Hinsman EJ (1973) Neuroglial response to sciatic neurectomy. I. Light microscopy and autoradiography. J Comp Neurol 151:237–254

Kershman J (1939) Genesis of microglia in the human brain. Arch Neurol Psychiatr 41:24–50

Killmann SA, Cronkite EP, Fliedner TM, Bond VP (1962) Cell proliferation in multiple myeloma studied with tritiated thymidine in vivo. Lab Invest 11:845–853

Kitamura T (1973) The origin of brain macrophages – some considerations on the microglia theory of Del Rio-Hortega. Acta Pathol Jpn 23:11–26

Kitamura T, Hattori H, Fujita S (1972) Autoradiographic studies on histogenesis of brain macrophages in the mouse. J Neuropathol Exp Neurol 31:502–518

Kitamura T, Tsuchihashi Y, Tatebe A, Fujita S (1977) Electron microscopic features of the resting microglia in the rabbit hippocampus identified by silver carbonate staining. Acta Neuropathol (Berl) 38:195–201

Kitamura T, Tsuchihashi Y, Fujita S (1978) Initial response of silver-impregnated "resting microglia" to stab wounding in rabbit hippocampus. Acta Neuropathol (Berl) 44:31–39

Knowles JF (1976) Cell death and the cell cycle in the subependymal layer of neonate rats after the administration of ethyl nitrosourea. Neuropathol Appl Neurobiol 2:365–376

Konyukhov BV, Sazhina MV (1971) Genetic control over the duration of G_1 phase. Experientia 27:970–971

Konyukhov BV, Sazhina MV (1976) The cell cycle and retinal histogenesis in fidget mutant mice. Dev Biol 54:13–22

Korr H (1973) Autoradiographische Untersuchungen zur Proliferation der Neuroglia im Hirn erwachsener Mäuse. Verh Dtsch Zool Ges 66:255–260

Korr H (1974) Autoradiographic studies on the cell cycle of glial cells in the adult mouse brain. In: Jilek L, Trojan S (eds) Ontogenesis of the brain, vol II. Universita Karlova, Praha, pp 461–466

Korr H (1978a) Autoradiographische Untersuchungen zur Proliferation verschiedener Zellelemente im Gehirn von Nagern während der prä- und postnatalen Ontogenese. Habilitationsschrift, University of Würzburg

Korr H (1978b) Combination of metallic impregnation and autoradiography of brain sections. A method for differentiation of proliferating glial cells in the brain of adult rats and mice. Histochemistry 59:111–116

Korr H (to be published) Autoradiographic studies of the proliferation of microglial cells and pericytes in the brain of untreated mice during the postnatal ontogeny

Korr H, Schultze B, Maurer W (1973) Autoradiographic investigations of glial proliferation in the brain of adult mice. I. The DNA synthesis phase of neuroglia and endothelial cells. J Comp Neurol 150:169–176

Korr H, Schultze B, Maurer W (1975) Autoradiographic investigations of glial proliferation in the brain of adult mice. II. Cycle time and mode of proliferation of neuroglia and endothelial cells. J Comp Neurol 160:477–490

Korr H, Knorre J, Schultze B, Maurer W (to be published a) Proliferation of endothelial cells in the brain of the young rat and adult mouse

Korr H, Schilling WD, Dauenhauer M, Schultze B, Maurer W (to be published b) Proliferation of glial cells in different areas of the forebrain of the 14-day old rat. I. Autoradiographic studies of cell cycle parameters

Korr H, Schilling WD, Dauenhauer M, Schultze B, Maurer W (to be published c) Proliferation of glial cells in different areas of the forebrain of the 14-day old rat. II. Autoradiographic studies of the mode of proliferation

Korr H, Schultze B, Maurer W (to be published d) Exchange of neuroglial cells between non-growth fraction and growth fraction in the brain of the adult mouse. Autoradiographic studies with ^{3}H- and ^{14}C-thymidine

Kraus-Ruppert R, Laissue J, Bürki H, Odartchenko N (1973) Proliferation and turnover of glial cells in the forebrain of young adult mice as studied by repeated injections of ^{3}H-thymidine over a prolonged period of time. J Comp Neurol 148:211–216

Kraus-Ruppert R, Laissue J, Bürki H, Odartchenko N (1975) Kinetic studies on glial, Schwann and capsular cells labeled with ^{3}H-thymidine in cerebrospinal tissue of young mice. J Neurol Sci 26:555–563

Kulenkampff H (1958) Untersuchungen zur Frage der Funktion des Ependyms im Zentralkanal des Rückenmarkes der erwachsenen weißen Maus. Z Anat Entwicklungsgesch 120:235–246

Kulenkampff H, Kolb W (1957) Mitosen im Ependym der erwachsenen weißen Maus. Naturwissenschaften 44:241

Lajtha LG (1963) On the concept of the cell cycle. J Cell Comp Physiol (Suppl 1) 62:143–145

Langman J, Welch GW (1967) Excess vitamin A and development of the cerebral cortex. J. Comp Neurol 131:15–26

Langman J, Shimada M, Haden C (1971) Formation and migration of neuroblasts. In: Cellular aspects of neural growth and differentiation. UCLA Forum Med Sci 14:33–59

Lapham LW, Lentz RD, Woodward DJ, Hoffer BJ, Herman CJ (1971) Postnatal development of tetraploid DNA content in the Purkinje neuron of the rat: an aspect of cellular differentiation. In: Cellular aspects of neural growth and differentiation. UCLA Forum Med Sci 14: 61–71

Lauder JM (1977) The effects of early hypo- and hyperthyroidism on the development of rat cerebellar cortex. III. Kinetics of cell proliferation in the external granular layer. Brain Res 126:31–51

Lawson SN, May MK, Williams TH (1977) Prenatal neurogenesis in the septal region of the rat. Brain Res 129:147–151

Leblond CP, Walker BE (1956) Renewal of cell populations. Physiol Rev 36:255–276

Lennartz KJ, Maurer W (1964) Autoradiographische Bestimmung der Dauer der DNS-Verdopplung und der Generationszeit beim Ehrlich-Ascitestumor der Maus durch Doppelmarkierung mit ^{14}C- und ^{3}H-Thymidin. Z Zellforsch 63:478–495

Lennartz KJ, Maurer W, Eder M (1968) Auswertungs-Verfahren bei Doppelmarkierung mit C-14- und H-3-Thymidin für exponentielles Wachstum. Histochemie 13:84–90

Lewis PD (1968a) A quantitative study of cell proliferation in the subependymal layer of the adult rat brain. Exp Neurol 20:203–207

Lewis PD (1968b) Radiosensitivity of the subependymal cell layer of the adult rat brain. Exp Neurol 20:208–214

Lewis PD (1968c) The fate of the subependymal cell in the adult rat brain, with a note on the origin of microglia. Brain 91:721–736

Lewis PD (1975) Cell death in the germinal layers of the postnatal rat brain. Neuropathol Appl Neurobiol 1:21–29

Lewis PD (1978) Kinetics of cell proliferation in the postnatal rat dentate gyrus. Neuropath Appl Neurobiol 4:191–195

Lewis PD (1979) The application of cell turnover studies to neuropathology. Recent Adv Neuropathol 1:41–65

Lewis PD, Lai M (1974) Cell generation in the subependymal layer of the rat brain during the early postnatal period. Brain Res 76:520–525

Lewis PD, Balazs R, Patel AJ, Johnson AL (1975) The effect of undernutrition in early life on cell generation in the rat brain. Brain Res 83:235–247

Lewis PD, Patel AJ, Johnson AL, Balázs R (1976) Effect of thyroid deficiency on cell acquisition in the postnatal rat brain: a quantitative histological study. Brain Res 104:49–62

Lewis PD, Fülöp Z, Hajós F, Balázs R, Woodhams PL (1977a) Neuroglia in the internal granular layer of the developing rat cerebellar cortex. Neuropathol Appl Neurobiol 3:183–190

Lewis PD, Patel AJ, Balázs (1977b) Effect of undernutrition on cell generation in the adult rat brain. Brain Res 138:511–519

Lewis PD, Patel AJ, Béndek G, Balázs (1977c) Effect of reserpine on cell proliferation in the developing rat brain: a quantitative histological study. Brain Res 129:299–308

Lima-de-Faria A (1965) Labeling of the cytoplasm and the meiotic chromosomes of Agapanthus with 3-H-thymidine, Hereditas 53:1–11

Ling EA, Leblond CP (1973) Investigation of glial cells in semithin sections. II. Variation with age in the numbers of the various glial cells types in rat cortex and corpus callosum. J Comp Neurol 149: 73–82

Ling EA, Paterson JA, Privat A, Mori S, Leblond CP (1973) Investigation of glial cells in semithin sections. I. Identification of glial cells in the brain of young rats. J Comp Neurol 149:43–72

Mann DMA, Yates PO (1973a) Polyploidy in the human nervous system. Part 1. The DNA content of neurones and glia of the cerebellum. J Neurol Sci 18:183–196

Mann DMA, Yates PO (1973b) Polyploidy in the human nervous system. Part 2. Studies of the glial cell populations of the Purkinje cell layer of the human cerebellum. J Neurol Sci 18: 197–205

Manuelidis L, Manuelidis EE (1971) An autoradiographic study of the proliferation and differentiation of glial cells in vitro. Acta Neuropathol (Berl) 18:193–213

Mares V (1975) An autoradiographic study of regional differences in DNA synthesis in the brains of young adult mice. Acta Histochem (Jena) 53:70–76

Mares V, Brückner G (1978) Postnatal formation of non-neuronal cells in the rat occipital cerebrum: an autoradiographic study of the time and space pattern of cell division. J Comp Neurol 177: 519–528

Mares V, Lodin Z (1970) The cellular kinetics of the developing mouse cerebellum. II. The function of the external granular layer in the process of gyrification. Brain Res 23:343–352

Mares V, Lodin Z (1974) An autoradiographic study of DNA synthesis in adolescent and adult mouse forebrain. Brain Res 76: 557–561

Mares V, Lodin Z, Srajer J (1970) The cellular kinetics of the developing mouse cerebellum. I. The generation cycle, growth fraction and rate of the external granular layer. Brain Res 23:323–342

Mares V, Schultze B, Maurer W (1974) Stability of DNA in Pukinje cell nuclei of the mouse. An autoradiographic study. J Cell Biol 63:665–674

Mares V, Lodin Z, Jilek M (1975) An estimate of the number of cells arising by division in mouse cerebral hemispheres from age one to 12 months: an autoradiographic study of DNA synthesis. J Comp Neurol 161:471–482

Matthews MA (1974) Microglia and reactive "M" cells of degenerating central nervous system: Does similar morphology and function imply a common origin? Cells Tissue Res 148:477–491

Maurer W, Schultze B (1968) Überblick über autoradiographische Methoden und Ergebnisse zur Bestimmung von Generationszeiten und Teilphasen von tierischen Zellen mit markiertem Thymidin. Acta Histochem [Suppl] (Berl) 8:73–87

Maurer W, Schultze B, Schmeer AC, Haack V (1972) Autoradiographic studies on the mode of growth in jenunal crypt cells of the mouse. J Microsc 96:181–189

McAllister JP II, Das GD (1977) Neurogenesis in the epithalamus, dorsal thalamus and ventral thalamus of the rat: an autoradiographic and cytological study. J Comp Neurol 172:647–686

McLardy T (1963) Thalamic microneurones. Nature 199: 820–821

Meller K, Breipohl W, Glees P (1966) Early cytological differentiation in the cerebral hemisphere of mice. An electron-microscopical study. Z Zellforsch 72:525–533

Mendelsohn ML (1960) The growth fraction: a new concept applied to tumors. Science 132:1496

Mendelsohn ML (1962) Autoradiographic analysis of cell proliferation in spontaneous breast cancer of C3H mouse. III. The growth fraction. J Natl Cancer Inst 28:1015–1029

Messier B, Leblond CP (1960) Cell proliferation and migration as revealed by radioautography after injection of thymidine-^{3}H into male rats and mice. Am J Anat 106:247–285

Messier B, Leblond CP, Smart I (1958) Presence of DNA synthesis and mitosis in the brain of young adult mice. Exp Cell Res 14:224–226

Miale IL, Sidman RL (1961) An autoradiographic analysis of histogenesis in the mouse cerebellum. Exp Neurol 4:277–296

Mitsui Y, Schneider EL (1976) Increased nuclear sizes in senescent human diploid fibroblast cultures. Exp Cell Res 100.147–152

Mori S (1972) Uptake of ^{3}H-thymidine by corpus callosum cells in rats following a stab wound of the brain. Brain Res 46:177–186

Mori S, Leblond CP (1969a) Identification of microglia in light and electron microscopy. J Comp Neurol 135:57–80

Mori S, Leblond CP (1969b) Electron microscopic features and proliferation of astrocytes in the corpus callosum of the rat. J Comp Neurol 137:197–226

Mori S, Leblond CP (1970) Electron microscopic identification of three classes of oligodendrocytes and a preliminary study of their proliferative activity in the corpus callosum of young rats. J Comp Neurol 139:1–30

Moskovkin GN (1976) Thyroid hormones in early postnatal development of the central nervous system: the effect of hyperthyroidism on the proliferate activity of the white matter cells in the rat cerebellum. Ontogenez 7: 350–354

Moskovkin GN, Fülöp Z, Hajós F (1978) Origin and proliferation of astroglia in the immature rat cerebellar cortex. A double label autoradiographic study. Acta Morphol Acad Sci Hung 26:101–106

Nachtwey DS, Cameron IL (1968) Cell cycle analysis. In: Prescott DM (ed) Methods in cell physiology, vol III. Academic Press, New York London, pp 213–259

Nass MMK (1969) Mitochondrial DNA: Advances, problems, and goals. Science 165:25–35

Nevmivaka GA (1964) A study of DNA synthesis and kinetics of the cell population during the development of the cerebellum cortex in white rate. In: Zhinkin LN, Zavarzin AA, (eds) A study of cell cycles and metabolism of nucleic acids during differentiation of the cells. Nauka, Moscow Leningrad, pp 107–115

Noetzel H (1962) Autoradiographische Untersuchungen über die physiologische Regeneration der Gliazellen. Verh Dtsch Ges Pathol 46:342–344

Noetzel H, Rox J (1964) Autoradiographische Untersuchungen über Zellteilung und Zellentwicklung im Gehirn der erwachsenen Maus und des erwachsenen Rhesus-Affen nach Injektion von radioaktivem Thymidin. Acta Neuropathol (Berl) 3:326–342

Nornes HO, Carry M (1978) Neurogenesis in spinal cord of mouse: an autoradiographic analysis. Brain Res 159:1–16

Oehmichen M (1975) Monocytic origin of microglial cells. In: Furth R van (ed) Mononuclear phagocytes in immunity, infection and pathology. Blackwell, Oxford, pp 223–240

Oehmichen M, Grüninger H (1974) Cytokinetic studies on the origin of cells of the cerebrospinal fluid. With a contribution on the cytogenesis of the leptomeningeal mesenchyme. J Neurol Sci 22:165–176

Oehmichen M, Grüninger H, Saebisch S, Narita Y (1973) Mikroglia and Pericyten als Transformationsformen der Blut-Monocyten mit erhaltener Proliferationsfähigkeit. Experimentelle autoradiographische und enzymhistochemische Untersuchungen am normalen und geschädigten Kaninchen- und Rattengehirn. Acta Neuropathol (Berl) 23:200–218

Oehmichen M. Grüninger H, Gencic M (1975) Experimental studies on kinetics and functions of monuclear phagozytes of the central nervous system. Acta Neuropathol [Suppl] (Berl) 6:285–290

Painter RB (1970) Repair of DNA in mammalian cells. Curr Top Radiat Res Quar 7:45–70

Pannese E, Ferrannini E (1967) Nuclear pyknosis in neuroglia cells of normal mammals. Acta Neuropathol (Berl) 8:309–319

Paterson JA, Leblond CP (1977) Increased proliferation of neuroglia and endothelial cells in the supraoptic nucleus and hypophysial neural lobe of young rats drinking hypertonic sodium chloride solution. J Comp Neurol 175:373–390

Paterson JA, Privat A, Ling EA, Leblond CP (1973) Investigation of glial cells in semithin sections. III. Transformation of subependymal cells into glial cells, as shown by radioautography after ^{3}H-thymidine injection into the lateral ventricle of the brain of young rats. J Comp Neurol 149:83–102

Pelc SR (1970) Metabolic DNA and the problem of ageing. Exp Gerontol 5:217-226

Pelc SR (1972) Metabolic DNA in ciliated protozoa, salivary gland chromosomes and mammalian cells. Int Rev Cytol 32:327–355

Pera F (1970) Mechanismen der Polyploidisierung und der somatischen Reduktion. Erg Anat Entwicklungsgesch 43:1–112

Peters A, Palay SL, de Webster FH (1970) The fine structure of the nervous system. The cells and their processes. Hoeber, Harper & Row, New York

Pilgrim C, Maurer W (1962) Autoradiographische Bestimmung der DNA-Verdopplungszeit verschiedener Zellarten von Maus und Ratte bei Doppelmarkierung mit ^{3}H- und ^{14}C-Thymidin. Naturwissenschaften 49:544–545

Pilgrim C, Lang W, Maurer W (1966) Autoradiographische Untersuchungen der Dauer der S-Phase und des Generationszyklus der Basalepithelien des Ohres der Maus. Exp Cell Res 44:129–138

Prestige MC (1974) Axon and cell numbers in the developing nervous system. Br Med Bull 30:107–111

Privat A (1975) Postnatal gliogenesis in mammalian brain. Int Rev Cytol 40:281–323

Privat A, Leblond CP (1972) The subependymal layer and neighboring region in the brain of the young rat. J Comp Neurol 146:277–302

Quastler H (1963) The analysis of cell population kinetics. In: Lamerton LF, Fry RJM (eds) Cell proliferation. Blackwell, Oxford, pp 18–34

Quastler H, Sherman FG (1959) Cell population kinetics in the intestinal epithelium of the mouse. Exp Cell Res 17:420–438

Raedler E, Raedler A (1978) Autoradiographic study of early neurogenesis in rat neocortex. Anat Embryol (Berl) 154:267–284

Rakic P, Sidman RL (1968) Subcommissural organ and adjacent ependyma: autoradiographic study of their origin in the mouse brain. Am J Anat 122:317–336

Reichard P, Estborn B (1951) Utilization of desoxyribosides in the synthesis of polynucleotides. J Biol Chem 188:839–846

Reinis S (1972) Autoradiographic study of ^{3}H-thymidine incorporation into brain DNA during learning. Physiol Chem Phys 4:391–397

Rickmann M, Wolff JR (1976a) Über die Entstehung von Astroblasten im Neocortex. Verh Anat Ges 70: 325–328

Rickmann M, Wolff JR (1976b) On the earliest stages of glial differentiation in the neocortex of rat. Exp Brain Res [Suppl] 1: 239–243

Rickmann M, Chronwall BM, Wolff JR (1977) On the development of non-pyramidal neurons and axons outside the cortical plate: the early marginal zone as a pallial Anlage. Anat Embryol (Berl) 151.285–307

Rio-Hortega P del (1921) Histogenesis y evolucion normal; exodo y distribucion regional de la microglia. Mem Real Soc Esp Hist Nat 11:213–268

Rio-Hortega P del (1932) Microglia. In: Penfield W (ed) Cytology and cellular pathology of the nervous system, vol 2. Hoeber, New York, pp 481–534

Roels H (1966) Metabolic DNA: A cytochemical study. Int Rev Cytol 19:1–34

Roessmann U, Friede RL (1968) Entry of labeled monocytic cells into nervous system. Acta Neuropathol 10:359–362

Sántha Kv (1932) Untersuchungen über die Entwicklung der Hortegaschen Mikroglia. Arch Psychiatr 96:36–67

Sántha Kv, Juba A (1933) Weitere Untersuchungen über die Entwicklung der Hortegaschen Mikroglia. Arch Psychiatr 98:598–613

Schaper A (1897a) The earliest differentiation in the central nervous system of vertebrates. Science 5:430–431

Schaper A (1897b) Die frühesten Differenzierungsvorgänge im Centralnervensystem. Roux' Arch Entwickl-Mech Org 5:81–132

Schlessinger AR, Cowan WM, Swanson LW (1978) The time of origin of neurons in ammon's horn and the associated retro-hippocampal fields. Anat Embryol (Berl) 154:153–173

Schultze B (1969) Autoradiography at the cellular level. In: Pollister AW (ed) Physical techniques in biological res., 2nd edn, vol III b. Academic Press, New York London

Schultze B, Oehlert W (1960) Autoradiographic investigation of incorporation of ^{3}H-thymidine into cells of the rat and mouse. Science 131:737–738

Schultze B, Gerhard H, Schump E, Maurer W (1973) Autoradiographische Untersuchung über die Proliferation der Hepatocyten bei der Regeneration der CCl$_4$-Leber der Maus. Virchows Archiv [Cell Pathol] 14:329–343

Schultze B, Nowak B, Maurer W (1974a) Cycle times of the neural epithelial cells of various types of neuron in the rat. An autoradiographic study. J Comp Neurol 158:207–218

Schultze B, Nowak B, Maurer W (1974b) Autoradiographic study of the proliferation of ganglionic cells during the embryonic development of the rat brain. In: Jilek L, Trojan S (eds) Ontogenesis of the brain, vol II. Universita Karlova, Prague, pp 467–475

Schultze B, Maurer W, Hagenbusch H (1976) A two emulsion autoradiographic technique and the discrimination of the three different types of labelling after double labelling with ^{3}H- and ^{14}C-thymidine. Cell Tissue Kinet 9:245–255

Schultze B, Kellerer AM, Maurer W (1979) Transit times through the cycle phases of jejunal crypt cells of the mouse. Analysis in terms of the mean values and the variances. Cell Tissue Kinet 12: 347–359

Schultze-Maurer B (1978) Cell proliferation and differentiation in the developing and adult mammalian brain. In: Laerum OD, Bigner DD, Rajewsky MF (eds) Biology of brain tumors. A series of workshops on the biology of human cancer, report No. 5, UICC technical report series, vol 30. Geneva, pp 39–59

Shimada M (1966) Cytokinetics and hostogenesis of early postnatal mouse brain as studied by ^{3}H-thymidine autoradiography. Arch Histol Jpn 26: 413–437

Shimada M, Yamano T, Nakamura T, Morikawa Y, Kusunoki T (1977) Effect of maternal malnutrition on matrix cell proliferation in the cerebrum of mouse embryo: an autoradiographic study. Pediatr Res 11:728–731

Sidman RL (1970) Autoradiographic methods and principles for study of the nervous system with thymidine ^{3}H. In: Nauta WJH, Ebbesson SOE (eds) Contemporary research methods in neuroanatomy. Springer, Berlin Heidelberg New York, pp 252–274

Sidman RL, Miale LL, Feder N (1959) Cell proliferation and migration in the primitive ependymal zone: an autoradiographic study of histogenesis in the nervous system. Exp Neurol 1:322–333

Silver J (1976) A study of ocular morphogenesis in the rat using ^{3}H-thymidine autoradiography: Evidence for thymidine recycling in the developing retina. Dev Biol 49:487–495

Sims TJ, Vaughn JE (1979) The generation of neurons involved in an early reflex pathway of embryonic mouse spinal cord. J Comp Neurol 183:707–720

Sinitsina VF (1971) DNA synthesis and kinetics of cellular populations in embryohistogenesis of mice. Arch Anat (Leningrad) 61:58–67

Skoff RP, Price DL, Stocks A (1976a) Electron microscopic autoradiographic studies of gliogenesis in rat optic nerve. I. Cell proliferation. J Comp Neurol 169:291–312

Skoff RP, Price DL, Stocks A (1976b) Electron microscopic studies of gliogenesis in rat optic nerve. II. Time of origin. J Comp Neurol 169:313–334

Smart I (1961) The subependymal layer of the mouse brain and its cell production as shown by radioautography after thymidine-^{3}H injection. J Comp Neurol 116:325–338

Smart I, Leblond CP (1961) Evidence for division and transformations of neuroglia cells in the mouse brain, as derived from radioautography after injection of thymidine-^{3}H. J Comp Neurol 116:349–367

Smith JA, Martin L (1973) Do cells cycle? Proc Natl Acad Sci USA 70: 1263–1267

Snell GD, Stevens LC (1966) Early embryology. In: Green EL (ed) Biology of the laboratory mouse, 2nd edn. McGraw-Hill, New York Toronto Sydney London, pp 205–245

Steel GG (1977) Growth kinetics of tumours. Oxford University Press, Oxford

Steel GG, Lamerton LF (1968) Cell population kinetics and chemotherapy. Natl Cancer Inst Monogr 30:29–50

Stensaas LJ (1975) Pericytes and perivascular microglial cells in the basal forebrain of the neonatal rabbit. Cell Tissue Res 158:517–541

Strauss BS (1968) DNA repair mechanisms and their relation to mutation and recombination. Curr Top Microbiol Immunol 44:1–85

Strauss BS (1974) Repair of DNA in mammalian cells. Life Sci 15:1685–1693

Sturrock RR (1974a) Histogenesis of the anterior limb of the anterior commissure of the mouse brain. I. A quantitative study of changes in the glial population with age. J Anat 117:17–25

Sturrock RR (1974b) Histogenesis of the anterior limb of the anterior commissure of the mouse brain. II A quantitative study of pre- and postnatal mitosis. J Anat 117:27–35

Sturrock RR (1974c) Histogenesis of the anterior limb of the anterior commissure of the mouse brain. III. An electron microscopic study of gliogenesis. J Anat 117:37–53

Sturrock RR (1974d) A light microscope study of glial necrosis with age in the anterior lime of the anterior commissure of the pre- and postnatal mouse. J Anat 117:469–474

Sturrock RR (1975) A light and electron microscopic study of proliferation and maturation of fibrous astrocytes in the optic nerve of the human embryo. J Anat 119:223–234

Sturrock RR (1977) Neurons in the mouse anterior commissure. A light microscopic, electron microscopic and autoradiographic study. J Anat 123:751–762

Sturrock RR (1978a) Development of the indusium griseum. II. A semithin light microscopic and electron microscopic study. J Anat 125:433–445

Sturrock RR (1978b) Development of the indusium griseum. III. An autoradiographic study of cell production. J Anat 126:1–6

Sturrock RR (1978c) A developmental study of epiplexus cells and supraependymal cells and their possible relationship to microglia. Neuropathol Appl Neurobiol 4:307–322

Swarz JR, Del Cerro M (1977) Lack of evidence for glial cells originating from the external granular layer in mouse cerebellum. J Neurocytol 6:241–250

Torvik A (1975) The relationship between microglia and brain macrophages. Experimental investigations. Acta Neuropathol [Suppl] (Berl) 6:297–300

Vaughan DW, Peters A (1974) Neuroglial cells in the cerebral cortex of rats from young adulthood to old age: an electron microscope study. J Neurocytol 3:405–429

Vaughn JE, Peters A (1971) The morphology and development of neuroglial cells. In: Cellular aspects of neural growth and differentiation. UCLA Forum Med Sci 14:103–140

Waechter Rv, Jaensch B (1972) Generation time of the matrix cells during embryonic brain development: an autoradiographic study in rats. Brain Res 46:235–250

Walker BE (1963) Infiltration and transformation of lymphoid cells in areas of spinal cord injury. Tex Rep Biol Med 21:615–630

Walker BE, Leblond CP(1958) Sites of nucleic acid synthesis in the mouse visualized by radioautography after administration of ^{14}C-labelled adenine and thymidine. Exp Cell Res 14: 510–531

Wallace RB, Altman J, Das GD (1969) An autoradiographic and morphological investigation of the postnatal development of the pineal body. Am J Anat 126:175–184

Wheeler KT, Lett JT (1972) Formation and rejoining of DNA strand breaks in irradiated neurons: in vivo. Radiat Res 52:59–67

Wilson CB, Hoshino T, Barker M, Downey R (1972) Kinetics of gliomas in rat and man. Progr Exp Tumor Res 17:363–372

Wimber DE, Quastler H (1963) A ^{14}C- and ^{3}H-thymidine double labeling technique in the study of cell proliferation in Tradescantia root tips. Exp Cell Res 30:8–22

Zavarzin AA, Stroyeva OG (1964) A study of DNA synthesis and kinetics of the cell population at differentiation of the retina and pigment epithelium and the iris by ^{3}H-thymidine-method. In: Zhinkin LN, Zavarzin AA (eds) A study of cell cycles and metabolism of nucleic acids during differentiation of the cells. Nauka, Moscow Leningrad, pp 116–125

Subject Index

Other Reviews of Interest in this Series

Part 6: **Lüdicke, M.**: Internal Ear
Angioarchitectonic of Serpents.
21 figures. 41 pages. 1978.
ISBN 3-540-08836-9

Volume 55

Part 1: **Reutter, K.**: Taste Organ
in the Bullhead (Teleostei).
20 figures. 98 pages. 1978.
ISBN 3-540-08880-6

Part 2: **Dvorák, M.**: The
Differentation of Rat Ova
During Cleavage. 62 figures.
131 pages. 1978.
ISBN 3-540-08983-7

Part 3: **Wagner, H.-J.**: Cell Types and
Connectivity Patterns in Mosaic Retinas.
30 figures. 81 pages. 1978.
ISBN 3-540-09013-4

Part 4: **Jones, D.G.**: Some Current
Concepts of Synaptic Organization.
21 figures. 69 pages. 1978.
ISBN 3-540-09011-8

Part 5: **Fleischer, G.**: Evolutionary
Principles of the Mammalian Middle
Ear. 25 figures. 70 pages. 1978.
ISBN 3-540-09140-8

Volume 56

Kaissling, B.; Kriz, W.:
Structural Analysis of the Rabbit Kidney.
47 figures. VIII, 123 pages. 1979.
ISBN 3-540-09145-9

Volume 57

Niimi, K., Matsuoka, H.:
Thalamocortical Organization of the
Auditory System in the Cat Studied by
Retrograde Axonal Transport of Horse-
radish Peroxidase. 30 figures.
X, 56 pages. 1979.
ISBN 3-540-09449-0

Volume 58

Verwoerd, C.D.A. van Oostrom, C.G.:
Cephalic Neural Crest and Placodes.
41 figures. VI, 75 pages. 1979.
ISBN 3-540-09608-6

Volume 59

Bär, T.: The Vascular System of the
Cerebral Cortex.
33 figures. VI, 60 pages. 1980.
ISBN 3-540-09652-3

Volume 60

Hildebrand, R.: Nuclear Volume
and Cellular Metabolism.
12 figures. VII, 54 pages. 1980.
ISBN 3-540-09796-1

Springer-Verlag Berlin Heidelberg New York